The Yellow Helmet

UNITED STATES NAVY

Blue Angels

FLIGHT HELMETS

HISTORY - DEVELOPMENT - EVOLUTION

Jeffrey D. Guidry

EGP

ISBN 978-1-943492-98-5

Elm Grove Publishing
San Antonio, Texas, USA
www.elmgrovepublishing.com

Table of Contents

	Foreword	4
	Introduction	7
CHAPTER 1	U.S. Navy Flight Exhibition Team	18
CHAPTER 2	AN-6540 Flight Helmet	26
CHAPTER 3	H-1 Flight Helmet	32
CHAPTER 4	H-2 Flight Helmet	51
CHAPTER 5	H-3 & H-4 Flight Helmets	59
CHAPTER 6	USAF P-1 Flight Helmet	67
CHAPTER 7	APH-5 Flight Helmet	76
CHAPTER 8	APH-6 Flight Helmet	97
CHAPTER 9	DH-151 Flight Helmet	116
CHAPTER 10	HGU-33/P Flight Helmet	141
CHAPTER 11	HGU-55/P Flight Helmet	151
CHAPTER 12	Newbies	171
CHAPTER 13	The Yellow Helmet	181
CHAPTER 14	VIP Flight Helmets	195
CHAPTER 15	Call Signs	204
CHAPTER 16	The Ground Crew	218
CHAPTER 17	Blue Angels Life Support	252
CHAPTER 18	Pensacola Naval Air Station	262
CHAPTER 19	GENTEX	268
CHAPTER 20	Protection, Incorporated	273
CHAPTER 21	Don't Judge a Book by its Cover	284
CHAPTER 22	The Blue Suit	306
CHAPTER 23	Trader Jon's	315
	In Memoriam	327
	Acknowledgements	329
	Index	330

Foreword

The Yellow Helmet is a fascinating book using the evolution of aviation head gear — from cloth and leather, to modern lightweight elements — as a parallel to the 75 years of the Blue Angels pursuit of excellence.

The detective talent of Jeff Guidry shines as he reveals reams of here-to-fore unseen photos of Blue Angels with diagrams of their protective equipment. By tying them together in a factual format, he discloses the personalities of the pilots who wore the helmets, as well as the talents of those who developed and maintained the most important piece of gear worn by the Blues.

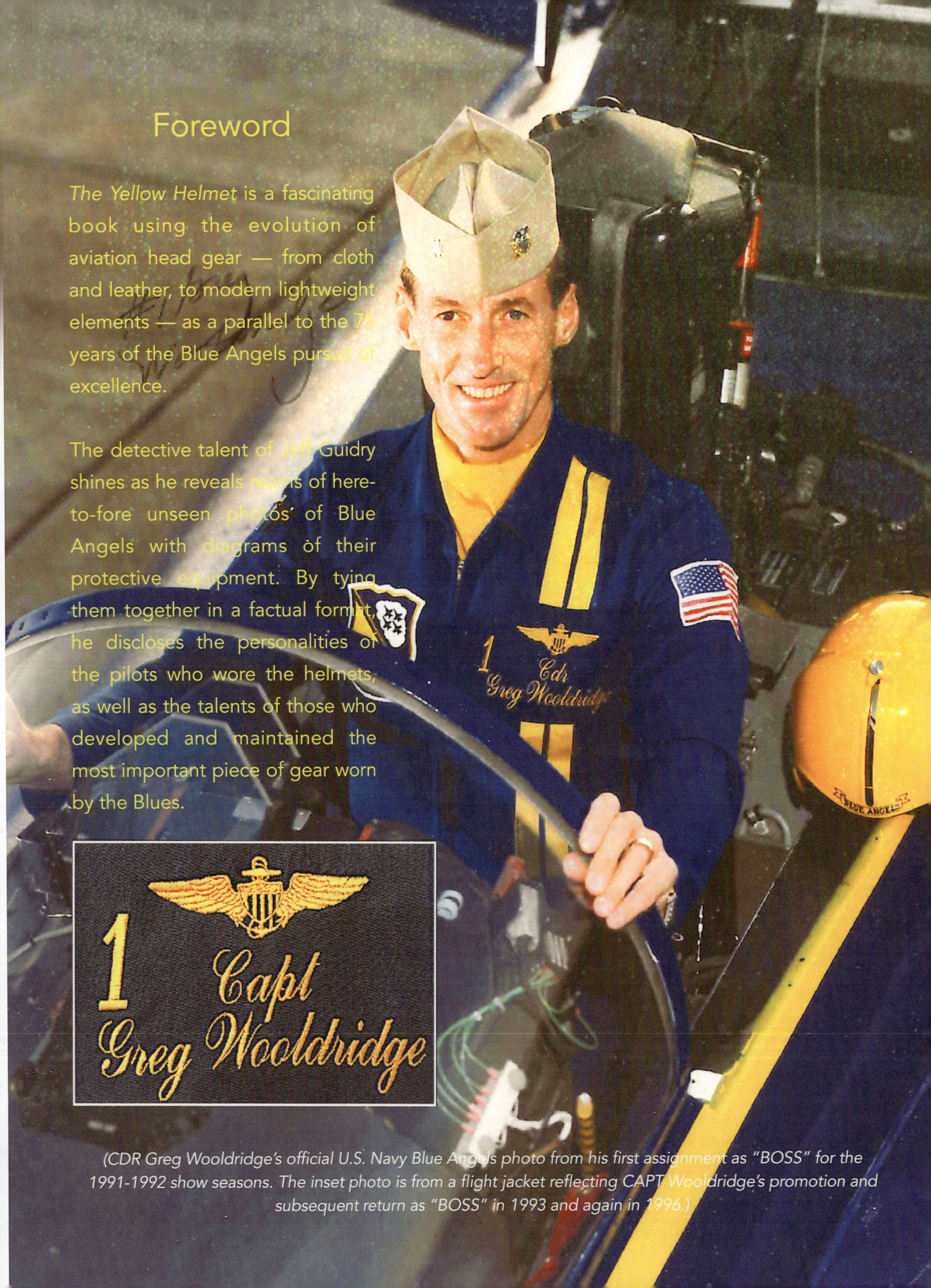

(CDR Greg Wooldridge's official U.S. Navy Blue Angels photo from his first assignment as "BOSS" for the 1991-1992 show seasons. The inset photo is from a flight jacket reflecting CAPT Wooldridge's promotion and subsequent return as "BOSS" in 1993 and again in 1996.)

I was captured by the way in which *The Yellow Helmet* shares stories and testimonials that I had not heard in my three stints leading the team as Boss in the #1 jet. Jeff's research revealed the changes in helmets necessitated by the team's progression from props to jets. The new, high-performance jets challenged the genius of the creators of protective helmets. Jeff digs deeply to show us this amazing transformation.

Jeff reminds us of the air show "unseens" who are the thunder to the lightning the team produces. Those men and women are often overlooked yet work tirelessly to ensure the safety of the pilots. Through real-life stories, he illustrates the commitment of devoted crew members who took personal pride in their contribution to the flight demonstration. That sense of ownership to exceptional outcomes can be felt throughout the book.

This is a book that is delightfully unique in its style. It matches technical gear elements to highly emotional and personal factors that create superior human performance. The iconic role the yellow helmet plays to the legend of the Blue Angels is made crystal clear. I thoroughly enjoyed reading *The Yellow Helmet* and I'm sure you will too!

Glad to be here!

Greg Wooldridge

Captain Greg "Rugdance" Wooldridge U.S. Navy (Retired)
Blue Angel #1: 1991-1992; 1993; 1996

Introduction

Happy 75th anniversary to the United States Navy Blue Angels!

The Blue Angels were born on 24 April, 1946, at NAS Jacksonville, Florida, and performed their first aerial demonstration on 15 June of the same year. The Team would relocate to NAS Corpus Christi, Texas and to NAS Whiting Field, Florida before finding their permanent home at NAS Pensacola in 1955.

According to their official Web site[1], "The mission of the Blue Angels is to showcase the pride and professionalism of the United States Navy and Marine Corps by inspiring a culture of excellence and service to country through flight demonstrations and community outreach."

Through the last 75 years, the Blue Angels have performed for an estimated 500 million spectators in every U.S. state and several foreign countries. Flight demonstration aircraft have included the F6F "Hellcat" (1946), F8F "Bearcat" (1946-1949), F9F-2 "Panther" (1949-1950), F9F-5 "Panther" (1951-1955), F9F-8 "Cougar" (1955-1957), F11F-1 "Tiger" (1957-1968), F-4J "Phantom II" (1969-1974), A-4F "Skyhawk II" (1974-1986), F/A-18A/B/C/D "Hornet" (1986-2020) and the F/A-18E/F "Super Hornet" (2021).

---

[1] https://www.blueangels.navy.mil/

The Yellow Helmet

Not long after writing my first book, *The Red Helmet - History, Development & Evolution of the USAF Thunderbirds Flight Helmet,*[2] I began fielding questions from my publisher, colleagues and friends in my flight gear collecting circles inquiring, "Have you given any consideration to writing a companion book about the Blue Angels flight helmets?" My initial response was, "One day at a time!"

Once the manuscript on the *The Red Helmet* was wrapping up, I decided to give this companion book some serious thought.

I started by reviewing research materials I have gathered over the past couple of decades, poured over hundreds of photographs and reached out to some other subject matter experts, including current and former Blue Angel contacts. The verdict was clear: It's a go!

My life-long love affair with aviation is undeniably my biggest vice — an obsession some would argue borders on the unhealthy. I learned to fly at 16 years of age; I built

[2] Elm Grove Publishing, ISBN 978-1-943492-76-3

an experimental airplane from a kit; I'm an admitted airshow fanatic; I own and operate a small, private aviation museum. Did I mention I love aviation?

This aviation passion — or obsession, depending on your perspective — has evolved into a fascination with military flight helmets, from their history, to collecting and restoration. Without a doubt, one of the most iconic, beautiful and interesting are the yellow helmets worn and flown by the U. S. Navy Blue Angels.

Top photo: 1998 Blue Angels Homecoming; Author, Renée (wife), sons Daniel, Chris and Stephen, and daughter Melissa.
Bottom photo: 1999 Blue Angels Homecoming: Author, Renée (wife), sons Chris and Stephen, and daughter Melissa.

I was first introduced to the Blue Angels in August of 1974. I was 16 years of age and a student pilot at Paul Fournet Air Service in Lafayette, LA. There was a "buzz" in the air (pun intended) about an air show at the former Chennault Air Force Base in Lake Charles, LA, about 75 miles west of Lafayette. While the details have faded with the passing of time, I distinctly remember that a group of pilots decided to make the trip. I was all in! This would be my first air show witnessing a U. S. military flight demonstration team. I could barely contain my excitement.

Following a series of accidents and the fuel crisis in 1973, the Blue Angels had just traded in their gas guzzling, twin engine F-4J "Phantoms" for the more fuel efficient, single engine A-4 "Skyhawk" also affectionately called the "Scooter."

Witnessing the Blue Angels for the first time defies description particularly to a young impressionable student pilot. What I didn't know at the time, was that this introduction to the Blue Angels would make a life-long impression on me.

Fast forward 23 years. My formal introduction to the Blue Angels came in 1997. I was working and living in New Orleans and in the early stages of my research on Blue Angels flight helmets. I made a "cold call" to the Blue Angels squadron using a phone number from their Web site. As you might imagine, the call was answered, "United States Navy Blue Angels" in a strong and confident voice. I asked to be transferred to the Life Support Shop. The voice at the other end answered, "Blue Angels Life Support Shop, This is Petty Officer DeRitter speaking, How can I assist you?" I learned that Petty Officer Brian DeRitter, an AME (AW)[3], was the supervisor in the Life Support Shop. I was confident I had been connected to the right person.

[3] The Aviation Structural Mechanic - Safety Equipment (AME), maintains and repairs utility systems throughout the aircraft; The AW represents "Enlisted Aviation Warfare Specialist"

United States Navy Blue Angels Flight Helmets

Within the Blue Angels organization, Life Support consists of seven aviation structural, mechanic, safety equipment and parachute rigger job specialists.

This shop maintains and repairs the systems that support the life of the pilot in normal and emergency situations. These systems include air conditioning, heating, avionics cooling, liquid oxygen, canopy and ejection seat emergency escape systems. Also maintained are the personal flight gear items that the pilot must wear such as helmets, oxygen masks, inflatable flotation devices, and the flight harness that attaches the pilot

to the ejection seat. The shop also sews flags and crests on Blue Angel uniforms and fabricates the protective covers for show aircraft.

I introduced myself and Brian's first question was, "Guidry? Where are you from?" I said I was calling from New Orleans but grew up near Lafayette, LA, to which Brian replied, "Wow, small world, I'm from New Iberia, LA!" If you don't have a map, New Iberia is about 15 miles south of Lafayette. Needless to say, I was off to a great start!

It didn't take long, or very much convincing, before I was offered, and gratefully accepted an invitation to visit NAS Pensacola. In November of 1997, my wife, Renée and I drove to Pensacola for the Blue Angels Homecoming Air Show. It was so good, we returned in 1998 and 1999 with the kids! And that was only the beginning of a life-long obsession with the Blue Angels and in particular, their end-of-season, Homecoming Air Show!

Although my first Blue Angels Homecoming was in 1996, my second visit in 1997 came with extra perks. I received a "tour" of the Blue Angels' offices and work areas, with some quality time in the Life Support Shop! This was the "inner sanctum" for anyone who collects military aviation flight gear! I was the proverbial kid in a candy store!

Our visit commenced on Saturday morning of the Homecoming Air Show weekend. The halls were bustling with blue uniforms, and you could feel the high energy in the air as these men and women made their final preparations for another flight demonstration. Despite the urgency of activity, everyone treated me like an honored guest. "Good morning, Sir!" was the automatic greeting from everyone I encountered in those hallowed halls. To say I was impressed would be a gross understatement.

Brian took such a personal interest in my flight helmet research that he graciously donated several authentic, show-flown helmet parts that had been removed from service due to age or wear, including a coveted gold visor. My Blue Angel "swag bag" contained some miscellaneous items and small flight clothing accessories. I still cherish this day despite the passing of more than two decades.

In 1998, Brian was serving as the Supervisor in Quality Assurance, which is comprised of a quality assurance officer and six maintenance technicians in various aviation ratings. This office prepares and assists in the preparation of maintenance instructions; reviews incoming technical directives and participates as members of the technical task force that investigates trouble areas and recommends corrective actions. Quality Assurance provides technical assistance to quality control inspectors and maintenance production personnel in the completion of aircraft maintenance, and regularly completes audits on work centers and specific maintenance programs.

That same year, Brian was selected as Sailor of the Year for his outstanding contribution to the Blue Angels. Impressive, but not surprising from a devoted sailor, husband, father and someone I am honored to call a friend.

In 1998 and 1999 we were invited to remain on base following the air show for a photo session with the Blue Angel pilots. This is typically reserved only for family members of Blue Angel personnel; however, Brian told us he would introduce us to the Team as "Uncle Jeff" and "Aunt Renée!" Having been born and raised in South Louisiana, I appreciated Brian's kind gesture, as it is common in Cajun country to refer to close friends as relatives, despite no known blood heritage. What a class act. I would later learn that Brian treated everyone with the same degree of kindness and compassion.

The year 1998 was also a banner year for me. Not only did I attend the Homecoming Air Show, but mid-season, the USAF Thunderbirds visited the Blue Angels at NAS Pensacola. As the story goes, every year, the Thunderbirds and Blue Angels visit their counterpart — one year at NAS Pensacola, or NAF El Centro, the other at Nellis AFB — to share best practices, encourage camaraderie between the demonstration teams, and perform a show for each other — back-to-back. Brian said these performances are closed to the public, and intended for base personnel and families only; however, since we were his "Aunt" and "Uncle" he could get us on base!

Detail of embroidered polo commemorating the 1998 reunion. (Author's collection)

Needless to say, this was, without a doubt, one of the most impressive air shows I have ever attended! To witness these two world class flight demonstration teams perform at the same venue, on the same day, is something you don't soon forget. I was fortunate to experience a second back-to-back performance in 2019 at Joint Base Andrews. Comparing and contrasting these two teams' performances was a dream come true! So, who's the best? That's very much a personal opinion! The shows are different, but at the end of the day, these two world-class air demonstration teams are both winners in my book!

The Blue Angels passion would last a lifetime. Brian would rotate out of the Blue Angels following the 1999 show season. He would remain in the Navy, complete OCS, commissioned a Warrant Officer (CWO3) and later retire with 22 years of service. We reunited with Brian, and his wife, Jill, in 2018 for the Blue Angels Homecoming Air Show at NAS Pensacola. Thanks for the memories, Sir! Let's not wait another 20 years to get together!

For the record, I am a student of aviation history, not an expert. My research for this book has been meticulous and I have endeavored to be as accurate as possible citing sources and photo credits. However, in many cases, the various elements necessary to corroborate a story or prove a claim are all too often vague or in some cases, non-existent. I have done my best to provide you the true story as I understand it offering the reader a balanced approach between authoritative and entertaining history. This is a reference work based on documented history and the oral testimonies of many others. I have made a conscious effort to bring the reader the most comprehensive information available. However, as this book was being published, new and additional information has been discovered. This is not unique to this work. There comes a time to draw the line, and that time was now. As I continue my pursuit of history, discover additional data, conduct additional interviews, or uncover historical documents once believed lost forever, perhaps a second addition will be in order.

It is worth mentioning that this book is not a comprehensive, detailed history of U. S. Navy flight helmets. These books already exist and were written by authors with far more knowledge on the subject than myself. I have elected to provide the reader with an "entry-level introduction" of each of the flight helmets used by the Team to provide some meaningful context, with a primary focus on the modifications, paint and custom graphics used by the Blue Angels over the past 75 years.

The Yellow Helmet

LCDR John "Gucci" Foley served with the Blue Angels as the Narrator (#7) in 1990, Opposing Solo (#6) in 1991 and Lead Solo (#5) in 1992. Now retired, he is the founder and president of John Foley, Inc., and is an industry leader in performance training to Fortune 500 corporations. His motivational program is based on the dynamics of debriefing — a key component to the Blue Angels' success.

> **"The Yellow Helmet. Icons are powerful tools. The Blue Angels have several icons, including the golden helmets worn by the team's six demonstration pilots. The helmets represent our quest for high performance and continuous improvement, our commitment to each other and the team, and our special responsibility to serve as ambassadors of good will to people all over the world."**
>
> LCDR John "Gucci" Foley
> U. S. Navy Blue Angels
> 1990 - 1992

One of LCDR Foley's topics is, "Glad to be Here: Lessons in High Performance From the Blue Angels."[4] Quoting from LCDR Foley's presentation, "The Yellow Helmet. Icons are powerful tools. They help organizations sustain a culture of shared values and beliefs. The Blue Angels have several icons, including the golden helmets worn by the team's six demonstration pilots. The helmets represent our quest for high performance and continuous improvement, our commitment to each other and the team, and our special responsibility to serve as ambassadors of good will to people all over the world."

[4] Not to be confused with the book, *Glad to be Here,* by former Blue Angel, Donnie Cochran who flew the Left Wing (#3) position for the 1986 and 1987 show seasons, the Slot (#4) position for the 1988 show season and returned as the "BOSS" (#1) for the 1995 and 1996 show seasons.

Powerful, meaningful words that embody who the Blue Angels are, and what they represent. LCDR Foley said of the yellow helmet, "So, I think this helmet really represents a culture." He also shared that when you are selected to the Team, you are not allowed to wear the Blue Angels crest until you earn the trust and respect of your teammates…every single day. "This helmet is an icon."

Icons: The Blue Angels crest, and the Yellow Helmet.

The stage was set. I was highly motivated and the timing also proved to be ideal. The year 2021 marks the 75th anniversary of the U. S. Navy Flight Exhibition Team, the Blue Angels! To commemorate this epic milestone the Blue Angels celebrated in grand style by trading in their F/A-18C & D "Hornets" for F/A-18E & F "Super Hornets!" The F/A-18 "Hornet" served the Blue Angels well for 35 years — since 1986 — when it replaced the A-4 "Skyhawk."

With a self-imposed deadline, I had no time to waste. I set my sights on Pensacola. My checklist was quickly growing: Coordinate with Blue Angels Public Affairs for a tour of the squadron, with an emphasis on the Life Support Shop; Interview Life Support personnel and photograph pilot and VIP flight helmets; Visit the National Museum of Naval Aviation for photographs; Visit the T. T. Wentworth Museum for photographs of the Trader Jon exhibit; Visit McGuires Irish Pub for photographs of their Blue Angels memorabilia…I'm sure I'm missing something!

And so, the journey and story of the yellow helmet begins.

Jeffrey D. Guidry

75TH
ANNIVERSARY

1
U.S. Navy Flight Exhibition Team

Following World War II the public interest in Naval Aviation quickly diminished. In response, Admiral Chester W. Nimitz, the Chief of Naval Operations, on April 24, 1946 ordered the formation of a flight demonstration team at Naval Air Station (NAS) Jacksonville, Florida. The Team's official name was the U. S. Navy Flight Exhibition Team, and they performed their first aerial demonstration on June 15, 1946 at their home base, NAS Jacksonville.

Rear Admiral Ralph Davison personally selected LCDR Roy M. "Butch" Voris to assemble and train a flight demonstration squadron, naming him Officer-in-Charge and Flight Leader. Voris selected LT Maurice "Wick" Wickendoll and LT Mel Cassidy, both veterans of the War in the Pacific, and fellow instructors to join him. They conducted their training in secret over the Florida Everglades so that, in Voris' words, "...if anything happened, just the alligators would know." A driving factor for the secrecy was concern that Army Air Corps legend, Jimmy Doolittle, now a general in the U.S. Army Air Forces, might just create a similar aerial demonstration team and beat the Navy in going public; however, It would be another 6 years before the Air Force created their own flight demonstration squadron, the Thunderbirds.

When the Blue Angels completed their training as a newly formed flight demonstration squadron, they announced via telegram to the Chief of Information, Washington, DC:

"Flight Exhibition Team is ready for future engagements. Team completed first public demonstration in F6F type aircraft at Southeastern Air Show, Jacksonville, on 15 and 16 June. Routine was received most enthusiastically. Team is an outstanding contribution for public information on Naval Aviation."

Not satisfied with "The Flight Exhibition Team," a contest was run throughout Training Command to find an appropriate name. A senior officer from Navy Headquarters proposed the name "Navy Blue Lancers" for consideration as a befitting moniker, but it was never fully accepted by the Team members. Voris once said, "The name 'Navy Blue Lancers' was about as exciting as kissing your sister!"

A newspaper reporter claimed to have pinned the name "Blue Angels" on the Team following an air show in Cleveland, Ohio, but Butch Voris tells us the way it really happened. While planning for an air show in New York, Lieutenant Maurice "Wick" Wickendol, was reading the *New Yorker* Magazine and saw an ad for a nightclub called the "Blue Angel." He suggested they use this name and the rest is history. The Team flew their first air show as the "Blue Angels" on July 19 - 21, 1946 in Omaha, Nebraska.

It would take another 3 years — May 1949 to be exact — before Flight Leader LCDR Raleigh "Dusty" Rhodes designed the first official Blue Angels insignia or "crest" (pictured above) and is nearly identical to the current design, 75 years later. Only the aircraft silhouettes in the cloud have changed with each aircraft the team has flown.

"The name 'Navy Blue Lancers' was about as exciting as kissing your sister."

LCDR Roy M. "Butch" Voris

The mission of the United States Navy Flight Demonstration Squadron is to showcase the pride and professionalism of the United States Navy and Marine Corps by inspiring a culture of excellence and service to country through flight demonstrations and community outreach.[1]

"I set the team up as all bachelors. I was a bachelor, so everybody's going to be a bachelor."

LCDR "Butch" Voris

In those formative years, the Blue Angels began their life as a demonstration squadron comprised of all bachelors. LCDR "Butch" Voris said, "I set the team up as all bachelors. I was a bachelor, so everybody's going to be a bachelor. We were not going to worry about children at this stage of the game."

SUPPER CLUBS

(Entertainment but no dancing.)

BLUE ANGEL, 152 E. 55 (PL 3-0626): Rose Murphy and Stuart Ross heading up a show that's perfectly all right for this lazy weather. . . . SPIVY'S ROOF, 139 E. 57 (EL 5-9215): Spivy's musings on the problems of fashionable life about town, Bart Howard's piano playing, and flamboyant music by the Three Flames. . . . RUBAN BLEU, 4 E. 56 (EL 5-9787): a typical Ruban Bleu mixture, but not quite as good as it usually is.

Advertisement for the Blue Angel supper club from *The New Yorker* magazine, June 15, 1946. (Author's personal collection)

As the Team upgraded their aircraft, their patches — or "crests" — received some minor upgrading as well, mainly so that the aircraft silhouettes in the cloud closely resembled their current show aircraft, and to commemorate significant Team anniversaries.

[1] https://www.blueangels.navy.mil/

United States Navy Blue Angels Flight Helmets

As you can imagine, whenever the Blue Angels upgraded their aircraft and their crests, their flight suits were undergoing a transition of their own. A few years following the inaugural Team in 1946, we began to see the Blue Angels step away from green and khaki U. S. Navy-issued flight suits to custom tailored blue and yellow flight suits.

Needless to say, with the periodic upgrading of show aircraft, crests and flight suits, it is reasonable that the Blue Angels' flight helmets would also be subject to regular scrutiny, and trading in the old for the new.

Similar to flight suits and crests, the flight helmets issued to and flown by Blue Angel pilots and crew members are flashy and aesthetically pleasing, but more importantly, it is a matter of life safety.

Since the Blue Angels' inception in 1946, the research and engineering around protective flight gear — specifically flight helmets — has made significant progress. In the chapters that follow, we witness the progression and transformation of the flight helmet from the days of thin, cloth caps that afforded the wearer with virtually no impact protection, to modern hard shell helmets built from fiberglass and Kevlar fabric impregnated with epoxy resin and molded under high heat and pressure.

These advanced helmets are equipped with custom fitted liners providing the wearer with impact protection as well as comfort. Advantages include high-tech communication components and superior sound attenuation in the high noise environment of a modern day jet fighter cockpit. Visors, visor housings and chin and nape straps work in concert to insure the helmet remains securely fastened even under the harshest possible circumstances: a high-speed ejection.

There is no denying that the Blue Angels' flight helmets have come a long way in the past 75 years, and deserve the continued attention to insuring this important piece of equipment is the best the U. S. Navy has to offer. Join me on a pictorial and historical journey that takes an in- depth examination of the development and evolution of the beautifully unique, and unquestionably stunning U.S. Navy Blue Angels Flight Helmets for the past 75 years.

I explore deep behind the scenes to capture how these helmets are manufactured, built to exacting standards, artistically painted, customized and coveted by each and every Blue Angel pilot.

Wanted — 8,000 "Blue Angels"

On the following page is an advertisement from *The Pittsburgh Press,* dated, Sunday, February 25, 1945, which reads:

"War casualties are being returned to this country at a rate of more than 30,000 a month. There is a shortage of personnel to aid in their care, treatment and rehabilitation. In progress is a campaign to procure 8,000 women to be trained as medical and surgical technicians for the Army's 60 general hospitals. Pittsburgh's quota is 200 a month. Candidates must have had two years of high school, be 20 to 50 years of age and have no dependents under 14. They should apply at the Women's Army Corps in the Jenkins Arcade. This "blue angel" is T 5 Mary E. Haumesser of Newark, O., surgical technician at Ashford General Hospital, White Sulphur Springs, W. Va. Women who sign up at the Pittsburgh office will be assigned to Ashford."

I was intrigued by this advertisement, in part because the use of the name "Blue Angels" predates the U. S. Navy Flight Demonstration Team by more than one year, and used by a sister organization, the U. S. Army!

ROTO
The Pittsburgh Press
Sunday, February 25, 1945

Wanted—8000 'Blue Angels'

War casualties are being returned to this country at the rate of more than 30,000 a month. There is a shortage of personnel to aid in their care, treatment and rehabilitation. In progress is a campaign to procure 8000 women to be trained as medical and surgical technicians for the Army's 60 general hospitals. Pittsburgh's quota is 200 a month. Candidates must have had two years of high school, be 20 to 50 years of age and have no dependents under 14. They should apply at the Women's Army Corps in the Jenkins Arcade. This "blue angel" is T 5 Mary E. Haumesser of Newark, O., surgical technician at Ashford General Hospital, White Sulphur Springs, W. Va. Women who sign up at the Pittsburgh office will be assigned to Ashford.

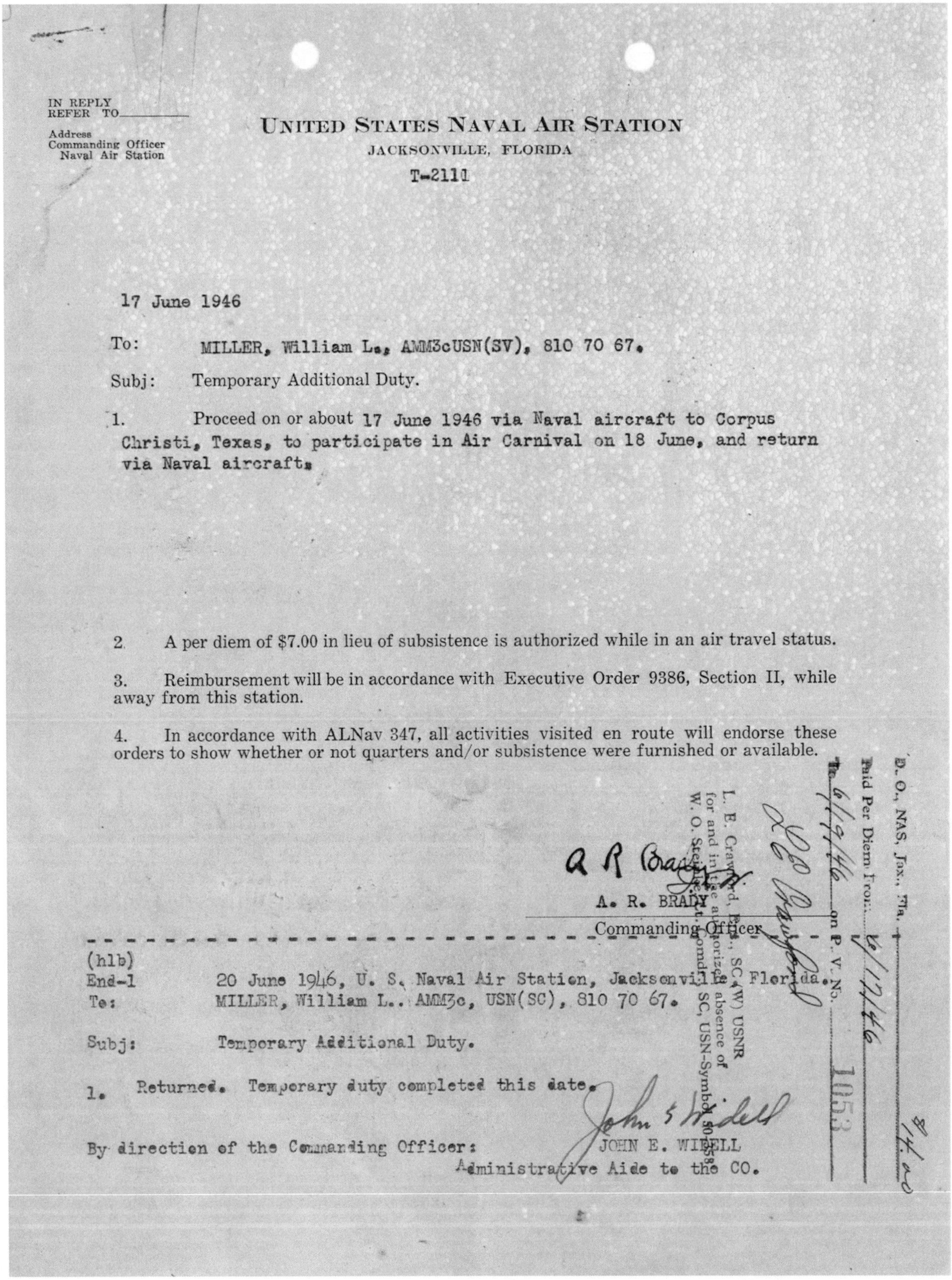

IN REPLY REFER TO
Address Commanding Officer Naval Air Station

UNITED STATES NAVAL AIR STATION
JACKSONVILLE, FLORIDA
T-2111

17 June 1946

To: MILLER, William L., AMM3cUSN(SV), 810 70 67.

Subj: Temporary Additional Duty.

1. Proceed on or about 17 June 1946 via Naval aircraft to Corpus Christi, Texas, to participate in Air Carnival on 18 June, and return via Naval aircraft.

2. A per diem of $7.00 in lieu of subsistence is authorized while in an air travel status.

3. Reimbursement will be in accordance with Executive Order 9386, Section II, while away from this station.

4. In accordance with ALNav 347, all activities visited en route will endorse these orders to show whether or not quarters and/or subsistence were furnished or available.

A. R. BRADY
Commanding Officer

D. O., NAS, Jax, Fla.
Paid Per Diem From 6/17/46 To 6/19/46 on P.V. No. 1053
L. E. Crawford, SC (W) USNR
for and in the absence of
W. O. Steinle, SC, USN-Symbol

(hlb)
End-1 20 June 1946, U. S. Naval Air Station, Jacksonville, Florida.
To: MILLER, William L., AMM3c, USN(SC), 810 70 67.

Subj: Temporary Additional Duty.

1. Returned. Temporary duty completed this date.

By direction of the Commanding Officer: JOHN E. WIDELL
Administrative Aide to the CO.

U.S. Navy orders, dated 17 June 1946 authorizing temporary duty support to the Blue Angels just 2 days following their inaugural air show.
(National Naval Aviation Museum archives)

BLUE ANGELS

2
AN-6540 Flight Helmet

In those first few formative years the Blue Angels did not give much, if any, attention to flashy flight suits and custom flight helmets. Instead, they wore U. S. Navy green nylon flight suits and flight helmets, the same as every other pilot in the fleet. It would take a few years before fashion would make its Team debut.

The most common flight helmet worn by the first Blue Angels pilots were constructed of lightweight, green nylon, manufactured by Slote & Klein, designated AN-6540; however, there are several reference photos showing Blue Angel pilots wearing different helmets, including the khaki cloth flight helmet, also manufactured by Slote & Klein and designated AN-6542. The latter example can be seen in the photo on the left of LCDR Raleigh Ernest "Dusty" Rhodes who served as the Blue Angels third Leader in 1947.

The photos on the next page show an early Blue Angels team in their Willis &

Geiger, Inc. lightweight nylon flight suits and their corresponding green nylon flight helmets by Slote & Klein. The two helmet photos on this page are from the author's private collection: Nylon AN-6540 shown on the bottom left, and a khaki cloth AN-6542 on the bottom right.

Fashion aside, nylon is not a "breathable" fabric and these helmets and suits were, without a doubt, uncomfortable in the warmer months. Ironically, they were designed for tropical wear! More importantly, nylon was not flame retardant and could melt on the wearer's skin if exposed to fire. Fortunately, the nylon flight gear was short lived as a stand alone, soft helmet. The nylon AN-6540 was used as a pattern on a later, 2-piece flight helmet that incorporated a soft, nylon inner helmet — not unlike the AN-6540 — paired with an outer hard shell helmet, designated by the U. S. Navy as and H-4, which will be discussed in detail in chapter 5.

The U. S. Navy photo on this page show the Blue Angel pilots in this early years wearing AN-6540 paired with black rubber B-8 or M-1944 goggles.

U. S. Navy photo.

These early cloth flight helmets of which there were many variations. They were designed to perform a specific function: A platform from which to attach goggles, communications and afford the pilot some degree of sound attenuation in a very noisy cockpit.

When that first Blue Angel team was in its infancy, the goal was focused on selecting and training the demonstration pilots, assembling a ground team, acquiring suitable show aircraft and developing a safe and entertaining show. Little thought — if any — went into what the pilots wore. However, within a few years, hard shell helmets made their debut and it didn't take long before U. S. Navy squadrons began to customize their helmets by applying unique decals, paint schemes and reflective tape. Decorative flight helmets became very popular during the Korean War, and the Blue Angels soon followed suit.

Join me on an historical and pictorial journey that takes an in depth examination of the development and evolution of the beautifully unique, and unquestionably stunning U. S. Navy Flight Helmets for the past 75 years. From the team's humble beginning in 1946 to the present day and beyond, I explore deep behind the scenes to capture how these helmets are manufactured, built to exacting standards, artistically painted, and customized.

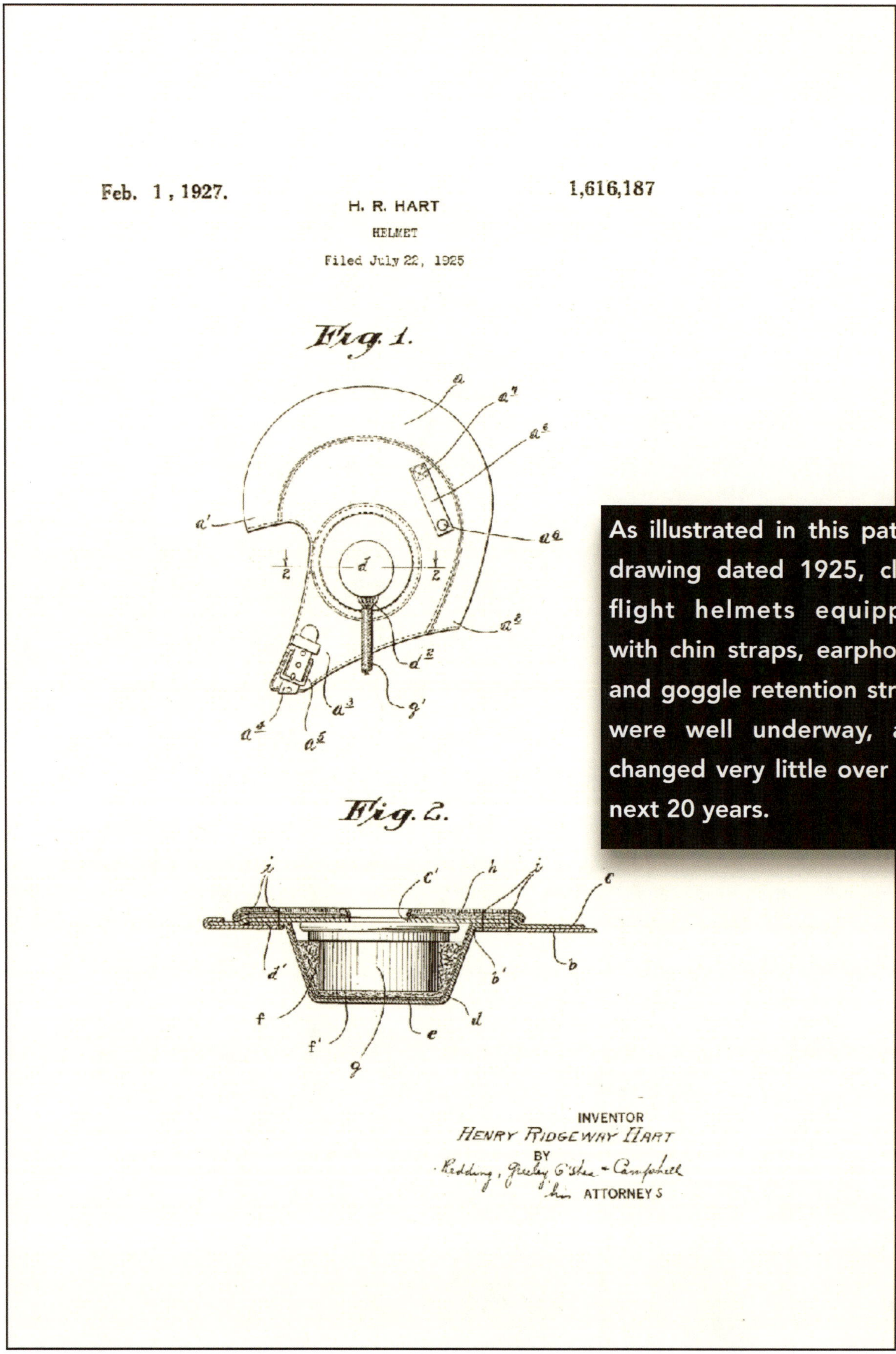

As illustrated in this patent drawing dated 1925, cloth flight helmets equipped with chin straps, earphones and goggle retention straps were well underway, and changed very little over the next 20 years.

AN-6540

3
H-1 Flight Helmet

General Textile Mills (GENTEX) produced its first hard shell helmet for the U. S. Navy in 1948 designated the H-1. This was a vast improvement over the leather and cloth helmets it replaced, which provided little, if any protective qualities. The Blue Angels would quickly adopt the new and improved flight helmets. However, they first flew with their H-1 helmets in standard factory white prior to painting them yellow and adding their iconic graphics.

Captain Roy Marlin "Butch" Voris was the first Blue Angels commander from 1946-1947, and returned as the commander in 1951.

(Voris Collection)

PROTECTIVE HELMET FOR HIGH SPEED AIRCRAFT

United States Patent Office

Patent No. 2,619,639

Edwin Hendler, Inventor

Filed December 1, 1949

This patent pertains to a protective helmet — in particular, a helmet used by pilots of "jet aircraft" — to protect the head from externally applied forces due to buffering, or crashing, as well as protection from low temperatures, wind blast and flash burn. This type aircraft intensifies the possible effect of head injuries that may occur during "accelerated takeoffs, violent acrobatic maneuvers, flights through turbulent air masses, and attempts at emergency escapes and crashes."

This helmet is described as a "Shell H" formed of a fiberglass base preferably laminated in about three complete layers and built up in thickness at the top area to about six layers. The fiberglass is to be laminated under pressure with polyester resin. The outer surface is to be finished with a sprayed coat of white lacquer. This external finish is of considerable importance in determining thermal protection afforded to the head of the wearer. The helmet offers excellent thermal protection relative to the high temperatures in the cockpits of jet planes.

Mounted on the inside of the helmet, by means of a suitable adhesive, is padding that is preferably formed of foam rubber, covered by a chamois skin. The padding is light weight, is capable of cushioning the head from blows and provides an elastic covering that readily follows the contours, while absorbing sweat.

A black rubber edge bead is cemented around the entire edge of the shell as a protection for the wearer and to aid in anchoring the helmet on the wearer. Male snaps are provided for attachment of an oxygen mask.

Dec. 2, 1952 **E. HENDLER** **2,619,639**

PROTECTIVE HELMET FOR HIGH-SPEED AIRCRAFT

Filed Dec. 1, 1949 2 SHEETS—SHEET 1

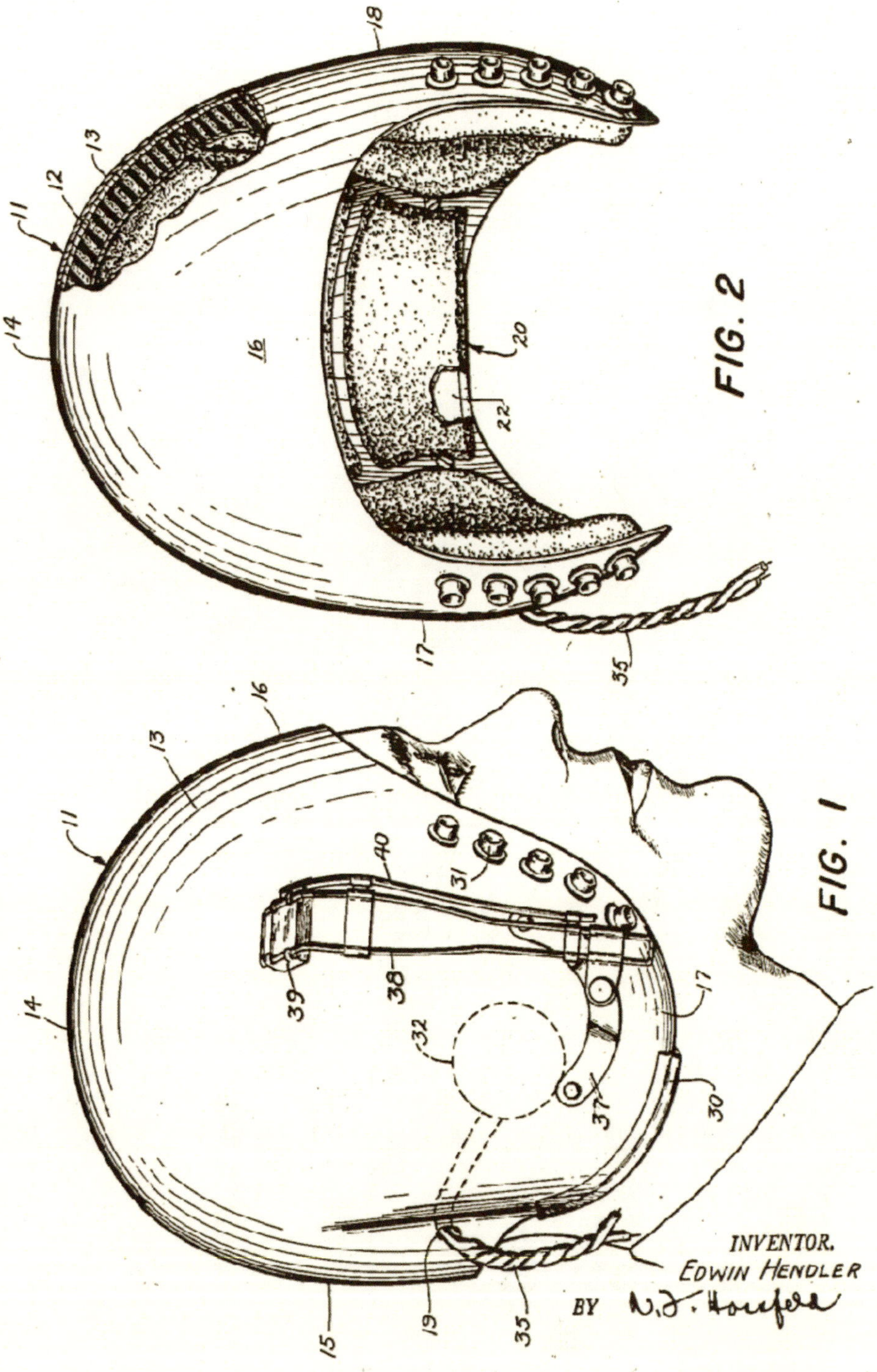

INVENTOR.
EDWIN HENDLER
BY N. J. Houfeld
ATTORNEY

Dec. 2, 1952 E. HENDLER 2,619,639

PROTECTIVE HELMET FOR HIGH-SPEED AIRCRAFT

Filed Dec. 1, 1949 2 SHEETS—SHEET 2

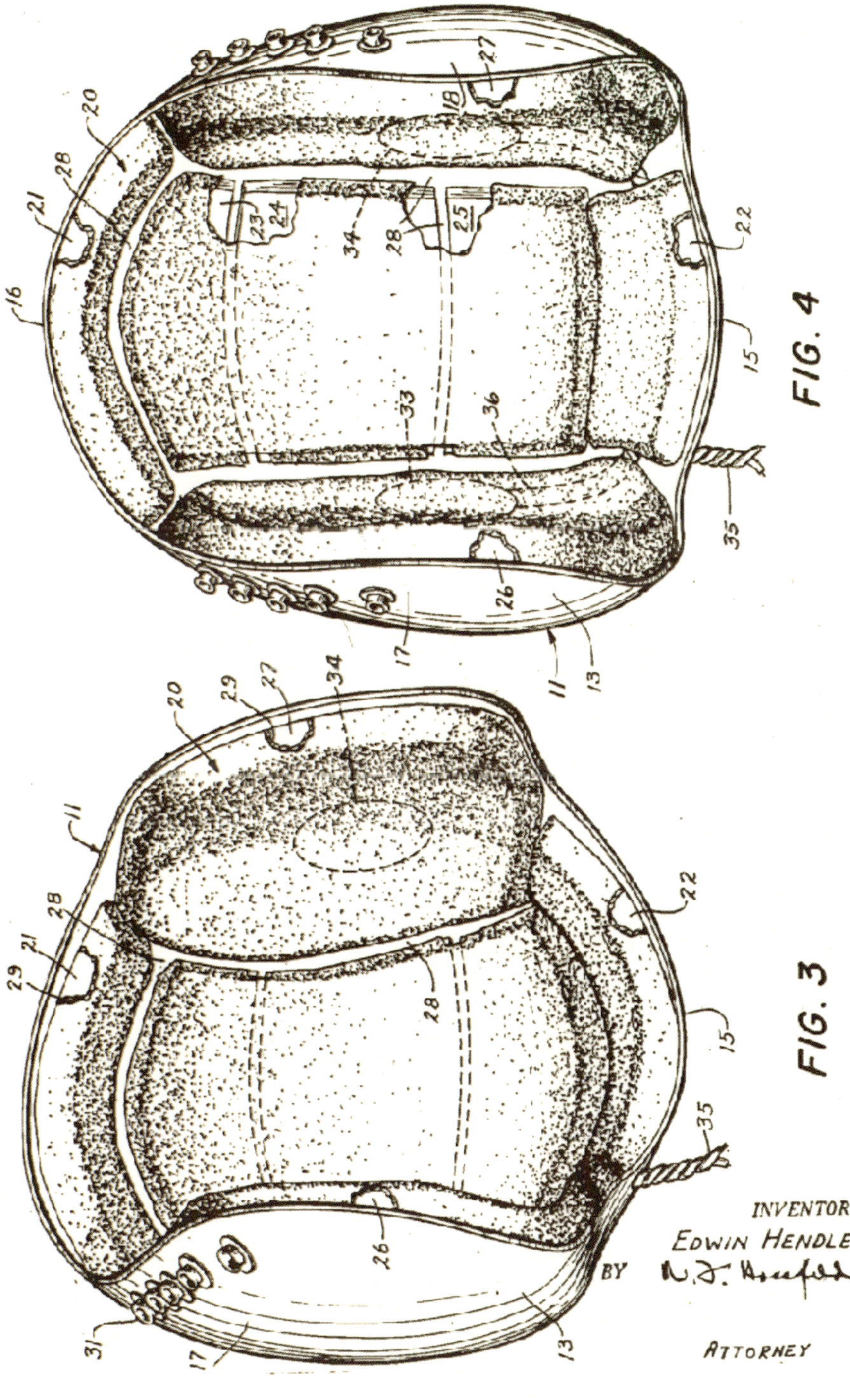

INVENTOR.
EDWIN HENDLER
BY N. J. Hoeufel
ATTORNEY

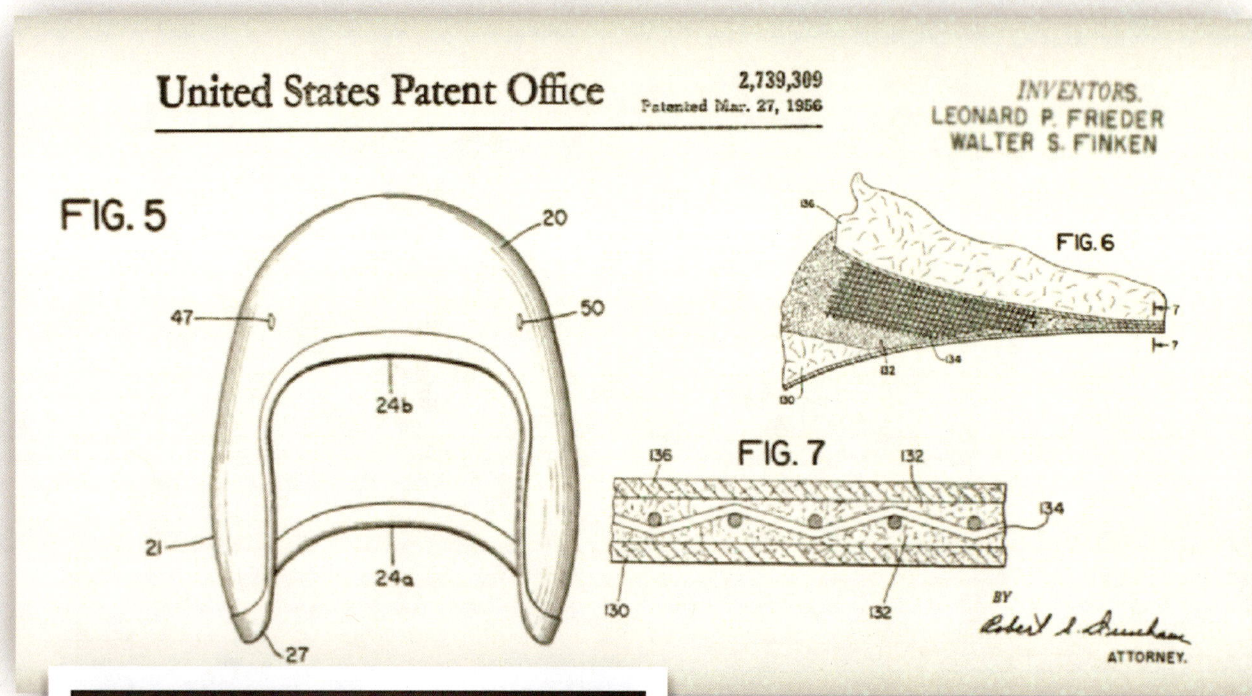

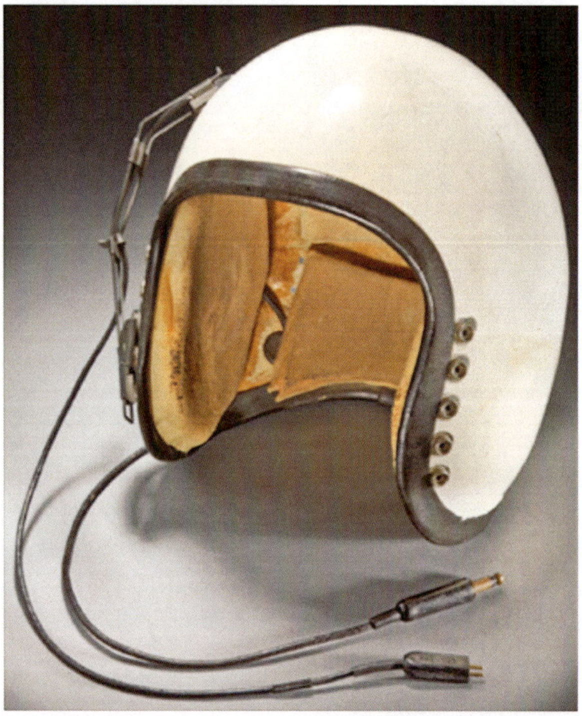

H-1 Flight Helmet, photo courtesy of http://www.salimbeti.com/

The early H-1 flight helmet was constructed of fiberglass cloth and resin, forming a hard protective shell to safeguard the crew member's head.

A boom mounted microphone was installed for operations below 10,000 feet, and a row of snaps on the left and right sides for attaching an oxygen mask for operations above 10,000 feet.

Later versions included a fixed chin strap, and added leather snap tabs riveted to the inside (left and right) of the helmet shell for attaching the oxygen mask to the helmet.

The photos on this page are from *LIFE Magazine*, 5 DEC 1949 (Author's collection) showing LCDR "Dusty" Rhodes (top left), LCDR John Magda (middle left), LT Arthur Hawkins (bottom left), and the rest of the Team wearing H-1 flight helmets. Although the photos are in black and white, it still appears that these H-1s were painted stock white and void of any Blue Angel graphics. That would soon change.

In this photo, date stamped 1949, the Team poses proudly in front of their newly acquired Grumman F9F-2 Panther. From left to right are Bob Longworth, Fritz Roth, Dusty Rhodes, George Hoskins, Ray Hawkins and Jake Robcke. The pilots appear to be using standard issue, white H-1 flight helmets.

Shortly thereafter, when aggression began to escalate in Korea, the Blue Angels were forced to temporarily disband in order to assist in the war effort. As a result, the Blue Angel pilots formed the nucleus of Fighter Squadron 191 (VF-191), "Satan's Kittens," under the command of World War II fighter ace and 1950 Blue Angels Commander/ Flight Leader, LCDR John Magda, who was killed in action on 8 March 1951.

In the Fall of 1951 the Blue Angels were reactivated and began performing for air show audiences again in May of 1952. LCDR Butch Voris was tasked with reorganizing the Team under his proven command and leadership.

With the formation of that new Team, we see the first examples of custom paint and graphics on their H-1 flight helmets. Although the photos are mostly in black and white it appears the helmets were painted yellow, with the U. S. Navy aviator wings decal above the brow, and the Blue Angels "banner" positioned below.

NAVAL AIR TRAINING COMMAND
Air Navy
Vol. 2, No. 8
SEPTEMBER
25c
U.S.N
Blue Angels Prepare for Combat
Painting by CDR Edward T. Grigware

During the Korean War, it became quite popular for Navy pilots to customize their flight helmets with bright colors, graphics and squadron influenced designs. Perhaps this was the impetus for the Blue Angels to follow suit upon their return. (U. S. Navy photos).

The second Blue Angels flight demonstration squadron to be commanded by LiCDR Roy M. "Butch" Voris (l to r): LT Tom Jones, LCDR R.M. Voris, LT Pat Murphy, LT A. R. Hawkins, LT Bud Rich, circa 1952.

All of the helmets in this photo are GENTEX H-1s that appear to be painted yellow as compared to their white socks and under shirts. The helmets also bear the U. S. Navy wings and Blue Angels "banner" on the front brow. There were no Blue Angels crests on the rear of the helmet during this time.

LT A. Ray "Hawk" Hawkins wearing a pristine H-1 flight helmet.
(U. S. Navy photo, circa 1952-1953)

A very young LT A. R. "Hawk" Hawkins during his first tour with the Blue Angels from 1948 to 1950. Seated proudly in his F-9F-2 "Panther" — the first jet used by the Blue Angels. LT Hawkins is wearing a white, standard U. S. Navy issued H-1 flight helmet, prior to switching to yellow paint and graphics.

More examples of H-1 helmets painted Blue Angel yellow. The photo on the right is LCDR Butch Voris (2nd tour as "Boss" - 1952), LT Ray Hawkins in the center (second tour, 1952 - 1953); Below is LCDR Frank Graham (1952 - 1953).

LT (jg) Dayle Crowe, 1954
(U. S. Navy photo)

The two photos on this page provide a view of the interior of the H-1 flight helmet. The chamois padded ear cups and liner appear to be gently used, evidenced by the lack of discoloration associated with perspiration stains.

Both H-1 flight helmets are a later version. The external oxygen mask snaps that were originally mounted to the left and right side of the helmet shell have been replaced by internally attached leather 3-snap tabs.

The Yellow Helmet

This stunning H-1 flight helmet is from the archives at the National Museum of Naval Aviation at NAS Pensacola. The curator, Dina Linn, was kind enough to grant me a research visit to capture this very early and rare example of a classic Blue Angel helmet. This particular helmet belonged to Captain Chuck Hiett the first U. S. Marine Corps pilot to join the Blue Angels in 1954. Since that time, a position on the Blue Angels has always been reserved for a USMC pilot. In a loving tribute to their USMC brother, the Blue Angels branded the back of his helmet "Jar-Head." (Author's photos)

2,619,639

PROTECTIVE HELMET FOR HIGH-SPEED AIRCRAFT

Edwin Hendler, Philadelphia, Pa.
Application December 1, 1949, Serial No. 130,576
3 Claims. (Cl. 2—6)
(Granted under the act of March 3, 1883, as amended April 30, 1928; 370 O. G. 757)

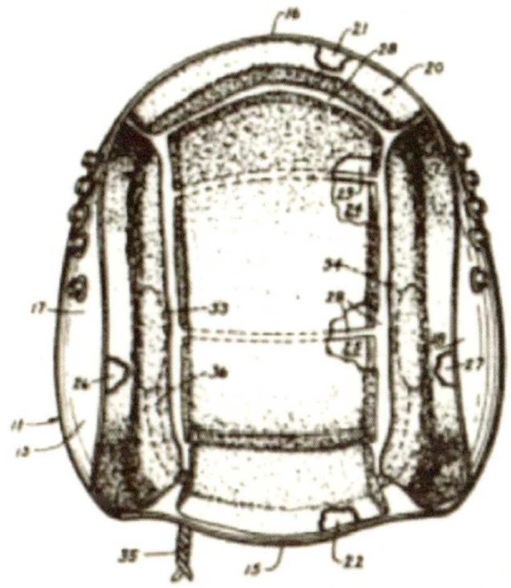

1. A helmet comprising a thin, light, rigid shell of laminated "fiberglas" shaped closely to the contour of a human head, said shell having an even outer surface substantially free from bulges, a smooth coating of white lacquer on the outside of said shell; a plurality of discrete foam rubber pads on the inside of said shell, including a pair of substantially oval pads adapted to cover a wearer's ears, a substantially rectangular front pad adapted to cover a wearer's forehead, a substantially rectangular top pad adapted to cover the top of a wearer's head, and a substantially rectangular back pad adapted to cover the back of a wearer's head, and a covering of chamois skin on the exposed surfaces of said pads, said pads being spaced from each other to form a plurality of intercommunicating ventilation channels.

Official Gazette of the United States Patent Office, Volume 665

This H-1 helmet photo, dated 1954, is a real beauty and depicts LCDR "Zeke" Cormier who served as the Blue Angels "Boss" for the 1954, 1955 and 1956 show seasons. (Official Photograph U. S. Navy, courtesy of the John Slezak collection).

Unidentified Blue Angels pilot wearing a GENTEX H-1 flight helmet.
(Photos courtesy of Richard Keane)

U.S.NAVY
U.S.NAVY
U.S.NAVY
U.S.NAVY
U.S.NAVY
U.S.NAVY
U.S.NAVY
U.S.NAVY

4
H-2 Flight Helmet

The H-2 protective flight helmet was an advanced version of the H-1 with the most noticeable difference being six reinforcing ridges radiating outward from a point at the crown of the helmet. These ridges intersect with a horizontal ridge that encircled the helmet shell just above the wearer's brow and ears. The H-2 was introduced in the early 1950s, and rather short lived before GENTEX introduced further variants designated the H-3 and H-4.

There are only a few examples of the Blue Angels using the H-2 flight helmet, and it didn't have a very long service life with the Team, or the U. S. Navy.

Photo credit: Smithsonian Air & Space Museum

On this page is a photo of CAPT Ray Hawkins, who flew with the Blue Angels from 1948 to 1950. He returned as the Commander for the 1952 and 1953 show seasons. In case you are not familiar with this man, I will enlighten you.

CAPT Arthur "Ray" Hawkins, one of the Navy's most decorated fighter pilots and was credited with shooting down 14 enemy planes by the age of 21. CAPT Hawkins succeeded Butch Voris following Voris' second tour at the helm of the Team.

He flew 142 combat missions, and was awarded 3 Navy Crosses, 3 Distinguished Flying Crosses, 4 Air Medals and a host of other awards. In addition, he was also credited with the sinking of the Japanese battleship *Ise.*[1]

[1] *Naval Aviation News,* April 1953.

On the previous page, Hawkins is holding the rare, short-lived GENTEX H-2 flight helmet, of which very few photos exist, and none known to exist in museums or private hands. The photo is from the personal collection of Jill Hawkins Votaw, daughter of CAPT Hawkins. A very similar photo of Hawkins with an H-2 helmet (slightly different pose and likely from the same photo session) resides at the T. T. Wentworth Museum in Pensacola, Florida and is from the estate of Jon "Trader Jon" Weissman.

CAPT Arthur "Ray" Hawkins, USN

The photo on the right captures CAPT Arthur "Ray" Hawkins wearing his GENTEX H-2 flight helmet.[2]

This impressive biography would not be complete without mentioning a story that first appeared in the *Saturday Evening Post* on April 13, 1954. I first read about this when the story was reprinted in the Fall 1992 issue of *Foundation* Magazine.

Written by CAPT Arthur Ray Hawkins, USN (Retired), he recounts the harrowing events of September 1953 when forced to eject from a factory fresh, Grumman F9F-6 "Cougar" while traveling faster than the speed of sound — and lived to tell the story!

[2] https://www.findagrave.com/memorial/35523099/arthur-ray-hawkins

The Blue Angels were flying the F9F-5 "Panther" when Grumman introduced the F9F-6 "Cougar." Of course, the Team wanted to upgrade, and several months later, they were awarded 6 new "Cougars" to be picked up from the Grumman factory in Long Island. They left their "Panthers" in Corpus Christi, Texas and boarded a Navy transport for the trip up north.

The day after their arrival, the pilots eagerly began their cockpit familiarization with their new "Cougars" to hone their cockpit resource management skills and insure a safe transition from the old to the new. Following a routine test hop, the aircraft performed as advertised and the Team departed for home, Corpus Christi. The flight proceeded without incident from Long Island to Sewart Air Force Base in Smyrna, Tennessee where they stopped for lunch and fuel. Upon departing Sewart, the Team made a 20-minute climb to 40,000 feet and leveled off, line abreast, to settle in for the 1-hour flight to Corpus Christi. Or so they thought!

CAPT Hawkins reported the flight controls were a little sloppy, but not uncommon at high altitudes. When the nose began to drop below the horizon, CAPT Hawkins instinctively applied aft pressure on the control stick to stop his descent — but the condition worsened. The aircraft controls continued to lose effectiveness, and soon he found himself in a 55 degree nose-down dive as the aircraft accelerated past the speed of sound!

"How tough is my helmet?"

CAPT Arthur Ray Hawkins

CAPT Hawkins was beginning to lose consciousness and knew instinctively that he had no choice but to eject before he was no longer capable of making the decision. However, the centrifugal forces on his body prevented him from reaching the necessary pre-ejection lever to initiate the explosive canopy release. Forced to resort to the emergency handle — only to be used in desperate cases — it would arm the ejection seat, but wouldn't blow off the canopy. This meant the ejection seat would launch CAPT Hawkins *through* the thick plexiglass canopy! CAPT Hawkins remembers asking himself, "How tough is my helmet?" He was about to find out!

CAPT Hawkins wrote, "When the ejection charge fired, I was four or five inches off the seat. It [ejection seat] came up and hit me like a pile driver. Too stunned to feel anything, I went through the canopy, a limp bundle traveling faster than the speed of sound. When the momentary blackout passed, I found myself clawing for my rip cord in a groggy attempt to open my parachute. Then, I thought, How stupid! Wait until you slow down. A chute opened at this speed would be torn to shreds. Then I realized I was bareheaded. The wind had torn away my face curtain, helmet and oxygen mask." (I can't help but wonder if that helmet was ever recovered on the ground.)

> **"Then I realized...the wind had torn away my face curtain, helmet and oxygen mask."**
>
> CAPT Arthur Ray Hawkins

Fortunately, the ejection was successful. Unfortunately, CAPT Hawkins was now hurling through the sky above 30,000 feet...with no oxygen. The plan at this point was to free fall until reaching an altitude with sufficient oxygen, but the hypoxia was already affecting his mental acuity. A blackout — total loss of consciousness — could be fatal. CAPT Hawkins played his only card and pulled his ripcord at an estimated 29,000 feet. The chute deployed, and CAPT Hawkins floated quietly under an undamaged canopy.

At this altitude, CAPT Hawkins remembers the air being very cold, and drifting in and out consciousness due to a lack of sufficient oxygen. As he slowly descended over Mississippi, the oxygen content increased and his vision and mental clarity were restored. He was now able to see a couple of his teammates circling him in their "Cougars" to keep a watchful eye on their "Boss." CAPT Hawkins wanted desperately to give a wave to indicate he was OK, but didn't have the strength due to his prolonged oxygen starvation. The descent from the bailout to reaching the ground lasted about a half an hour, but seemed longer.

Once reaching 10,000 feet, breathing returned to normal, and by 5,000 feet, CAPT Hawkins was able to summon enough strength to raise his arms and signal that he was

OK. CAPT Hawkins landed in a cotton field and was immediately tended to by some people in a pickup truck who saw him land. Miraculously, Capt. Hawkins only suffered a bruised rib and a few scratches, to which he commented, "For a man who had been through a supersonic bailout, I was in pretty good shape." That's an understatement, Sir!

"For a man who had been through a supersonic bailout, I was in pretty good shape."

CAPT Arthur Ray Hawkins

An accident review board determined that CAPT Hawkins' aircraft suffered a malfunction of the "flying tail" as a result of a slow hydraulic leak. The other 5 aircraft were checked and cleared to fly, and CAPT Hawkins ordered a replacement "Cougar" without delay.

In the photo on the right (circa 1957), LT Bruce Bagwell (right) confers with an unidentified person who is wearing an H-2 flight helmet. Even though the photo is black and white, it appears the helmet is painted yellow based on its similarity to the Blue Angels "script" on the aircraft, which is known to be yellow, and noticeably darker than the white clouds on the crest. (Photo courtesy of John Slezak).

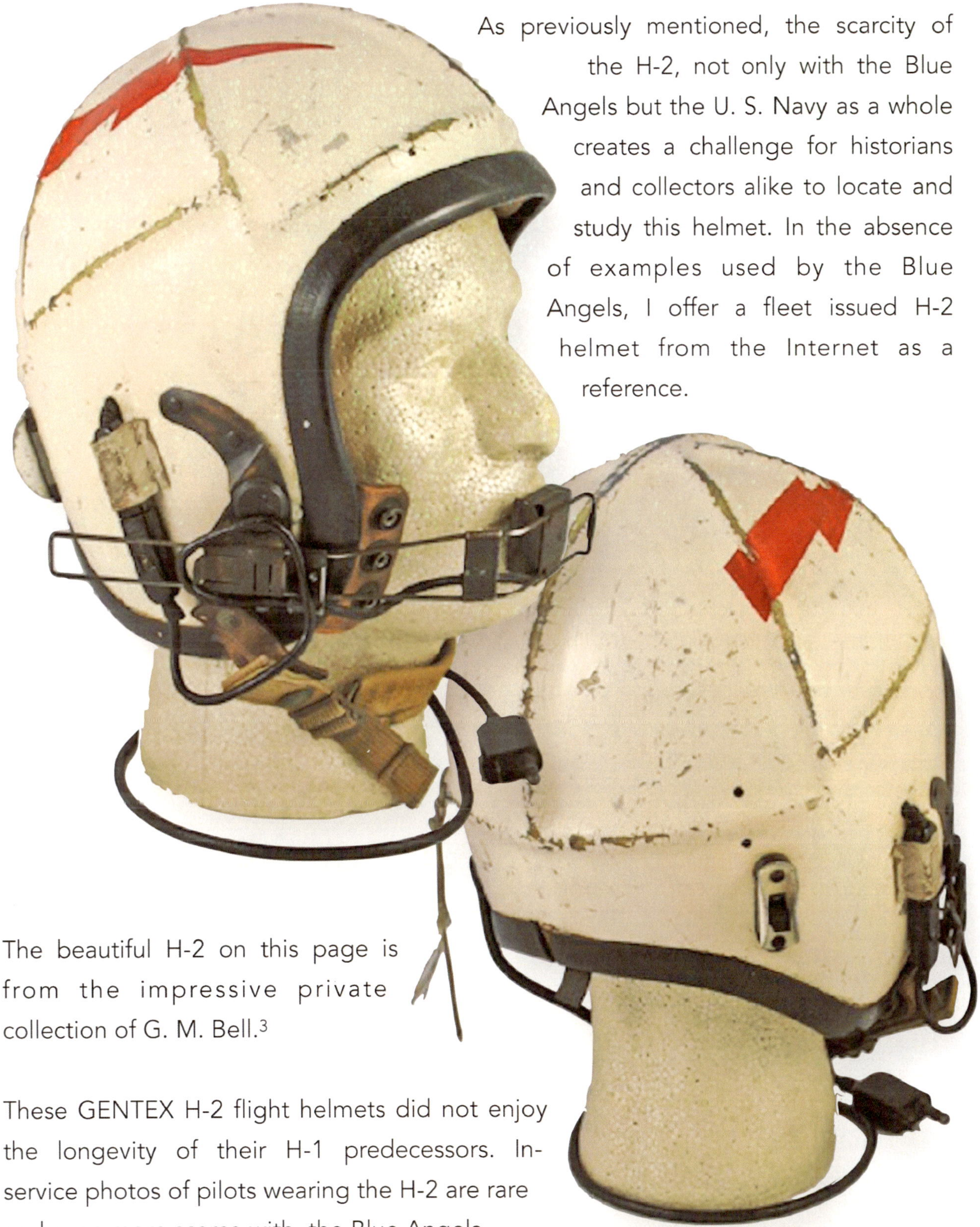

As previously mentioned, the scarcity of the H-2, not only with the Blue Angels but the U. S. Navy as a whole creates a challenge for historians and collectors alike to locate and study this helmet. In the absence of examples used by the Blue Angels, I offer a fleet issued H-2 helmet from the Internet as a reference.

The beautiful H-2 on this page is from the impressive private collection of G. M. Bell.[3]

These GENTEX H-2 flight helmets did not enjoy the longevity of their H-1 predecessors. In-service photos of pilots wearing the H-2 are rare and even more scarce with the Blue Angels.

---

3 bellsaviation.com and flighthelmet.com.

As the years pass, these helmets become increasingly scarce, underscoring the importance of preserving historically significant examples. As historians and in particular collectors, we should strive to be responsible custodians of these rare artifacts, lest they all vanish.

We should also be cognizant of the fact that we are only *temporary* custodians, and should strive to foster and encourage the next generation of care takers ensuring these beautiful artifacts remain a tangible part of our rich aviation heritage and history.

In Chapter 5, we explore GENTEX's next generation of flight helmets, the model H-3 and model H-4. Interestingly, GENTEX enjoyed a respectable business with the U. S. Navy and their latest H-series flight helmets, but not so much with the Blue Angels.

(Photo credit: *Foundation* Magazine, Volume 13, Number 2, Fall 1992)

5
H-3 & H-4 Flight Helmets

According to the NAVPERS (Navy Personnel) 10358-A, Parachute Rigger 3 & 2 NAVY Training Course[1], the H-3 protective helmet consists of a cloth under-helmet (liner) and a rigid protective shell. As shown in figure 14-1, the liner fits snugly, provides an attachment for the oxygen mask, and includes a set of radio communication earphones. Figure 14-2 shows an inside view of the protective shell. Figure 14-3 illustrates the fully equipped H-3 protective helmet assembly.

The liner is available in five sizes: small, medium, intermediate, large and extra large. The size of the liner can be adjusted by means of a nylon strap and buckle that is sewn on the back of the liner.

The oxygen mask is attached to the cheek flap by means of two Pull-the-Dot snap fasteners. The matching female fasteners on the mask have a dot on their outer surface, and unlock in one direction when pulled apart away from the dot. This provides positive locking action, insuring the pilot of keeping the oxygen mask even if he loses the protective helmet during a high-altitude bailout.

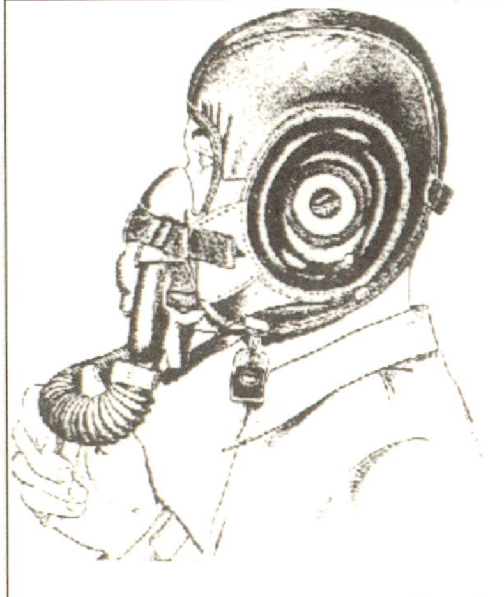

Figure 14-1. —Cloth underhelmet (liner).

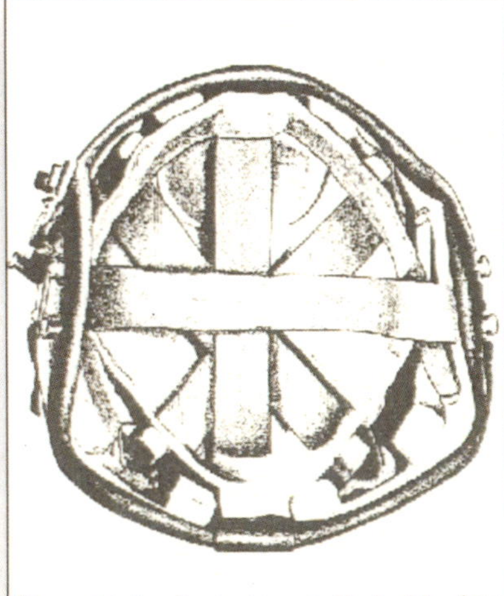

Figure 14-2. —Protective shell, inside view.

Figure 14-3. —Fully equipped H-3 protective helmet.

[1] http://www.best-of-flightgear.dk/

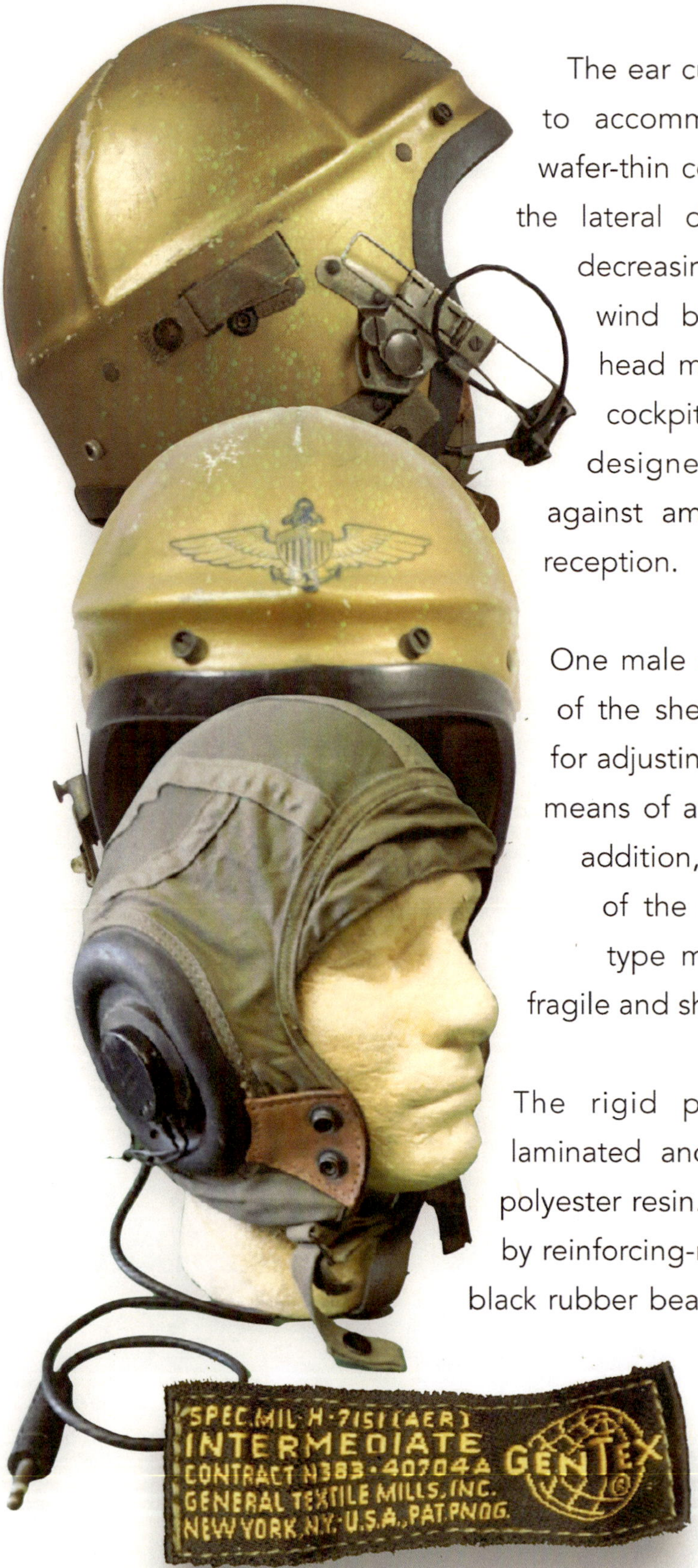

The ear cups on the liner have been designed to accommodate wafer type receivers. The wafer-thin construction of the receiver decreases the lateral dimensions of the helmet, thereby decreasing the frontal area exposed to the wind blast during bailout and facilitating head movements within the confines of the cockpit canopy. The earcups have been designed to provide sound attenuation against ambient noise, and to increase radio reception.

One male snap fastener is locate on each side of the shell exterior between the two buckles for adjusting the headband. This also provides a means of attaching the goggles to the shell. In addition, a fitting is secured to the right side of the shell exterior for attaching a boom type microphone. This microphone is very fragile and should be handled with care.

The rigid protective shell is composed of laminated and mat fiberglass compounded with polyester resin. The rigidity of the shell is increased by reinforcing-ridges located on its outer surface. A black rubber beading around the borders of the shell protects the wearer from sharp edges and provides a decorative finish.

The protective shell is available in four sizes; small, medium, large, and extra large. Tests have shown that the Navy's protective helmets will withstand an impact force in excess of 64 foot-pounds and a static load in excess of 600 pounds without structural failure.

WARNING: The wearer of the protective helmet should be instructed to not place any metal objects on the shell, especially above ear level. Under crash conditions, these additional metals may be driven into the head as secondary missiles.

Additional protection against injury from impact is provided by energy-absorbing padding inside the helmet shell, consisting of four pieces of Royalite cemented directly to the shell interior and small pieces of leather-covered foam rubber placed circumferentially between the adjustable headband and the shell. Deformation of all parts of the padding under impact absorbs the impact energy, thereby decreasing the injury potential of a blow delivered to the helmet shell.

The shell is fitted to the head by means of a sling suspension having an adjustable headband. Buckles for adjustment are located on both sides of the shell exterior just above the ear level. The most protection against impact is obtained when the headband is adjusted to fit snugly. Pilots and aircrewmen should be instructed in the proper procedure to adjust the slings very carefully prior to use.

A modification to the H-3 consists of installing stabilizing straps and chinstrap to provide the additional stability needed when catapulting, on arrested landings, and in maneuvers involving high "g" accelerations.

The wearer need not separate the assembly after it is modified since it can be utilized as a one-piece helmet. The modified cloth under-helmet (liner) should not be worn by itself since attempts to anchor the webbing by additional fasteners on the liner may increase the possibility of serious head injury in emergencies.

The H-3 helmet had become standard U. S. Navy issue to flight crews by April 1951. The outer shell was initially not attached to the inner liner, and was designed, allowing it to be carried away by the wind blast from an ejection. Later examples were modified with a series of straps on the inner liner that connected to snaps on the rigid helmet.

Conventional thinking suggested it would be prudent to maintain some degree of head protection resulting from an emergency egress, parachute descent, and landing.

The inner liner would remain with the pilot. The gold paint was also standard for the time, the thought being it provided some degree of thermal protection from solar radiation.

So, the question is: Did the Blue Angels use the H-3 or H-4 flight helmets?

The answer to this question remains...complicated. While there is very little evidence to go on, a few photos have surfaced. The photo above is from 1957 and shows Blue Angel pilot LT Mark Perrault in an H-3 or H-4 helmet, while his VIP back-seater sports an APH-5 helmet (discussed in Chapter 7).

Although not a show aircraft, the Blue Angels used a T2V "Seastar" as a support aircraft, to include VIP, media and incentive rides. Bob Moore, a Blue Angel AM3 (1952-1953) sent me a photo of a T2V he painted while on the Team (next page). It is interesting to note the pilot and person in the back seat are wearing what appears to be H-3 flight helmets, based on the radial "ribs" along the crown, the two-piece hard shell and cloth liner, and the suspension harness adjustment strap on the pilot's helmet.

The photo immediately above, from the *New Orleans States* newspaper, shows reporter Emmet Collins (left) give a thumbs-up before his VIP ride courtesy of LT Bruce Bagwell prior to a 1956 air show at the Naval Air Station in Belle Chasse, Louisiana. Collins is wearing an H-2 helmet, while Bagwell is wearing an H-4 helmet. (Photo courtesy of John Slezak)

Original H-3 instruction sheet from General Textile Mills (now GENTEX), 1952. Credit: http://www.best-of-flightgear.dk/h3helmet.htm.

INSTRUCTIONS

PILOT'S PROTECTIVE HELMET TYPE H-3

Made in five (5) head sizes -- Small, Medium, Intermediate, Large, and Extra Large

Nominal equivalent head sizes:

ASSEMBLY STOCK NO.	SIZE	SIZE RANGE	OUTER HELMET STOCK NO.
R37-H-1832-130	Small	6-5/8 to 6-7/8	R37-H-1840-30
R37-H-1832-125	Medium	7 to 7-1/8	R37-H-1840-20
R37-H-1832-120	Intermediate	7-1/8 to 7-2/8	
R37-H-1832-110	Large	7-2/8 to 7-3/8	R37-H-1840-10
R37-H-1832-105	Extra Large	7-3/8 to 7-5/8	R37-H-1840-5

Select the Protective Helmet most nearly your head size.

This unit is made to give you the maximum protection and comfort.

ADJUSTMENT AND USE:

1. Place the Cloth Flying Helmet on your head, and adjust it to your comfort as indicated on the instruction sheet supplied with that helmet.

2. Unlock the four (4) side adjusting buckles on the outside of your Protective Helmet, and slacken the web so that it will be larger than your head size.

3. Place the Protective Helmet on your head over the Cloth Flying Helmet, and draw the four (4) web ends equally to a comfortably taut position.

4. Remove the Protective Helmet from your head and draw the four (4) ends a little tighter to compensate for adjusting the lock of the tabs. Lock the web ends in position. The sweatband suspension system should now fit your head in a comfortable yet taut position.

5. The goggles are to be attached to this Protective Helmet with fasteners located between the adjusting buckles.

6. This Protective Helmet is attached to the Cloth Flying Helmet with a break link. If you are required to bail out at high altitude, and lose this helmet, you will still have the oxygen mask which is attached to the cloth flying helmet used with this unit.

The suspension system used in this helmet has been designed to:

(1) Firmly support your head at a distance from the shell of the Protective Helmet, so that if you do receive a blow you will at no time come in contact with the hard shell.
(2) The flexible sweatband will adjust itself to your own head contour.
(3) The crown suspension bands are fixed in position and will self-adjust to the shape of your head over the crown.

GENERAL TEXTILE MILLS, INC.
CARBONDALE, PA.

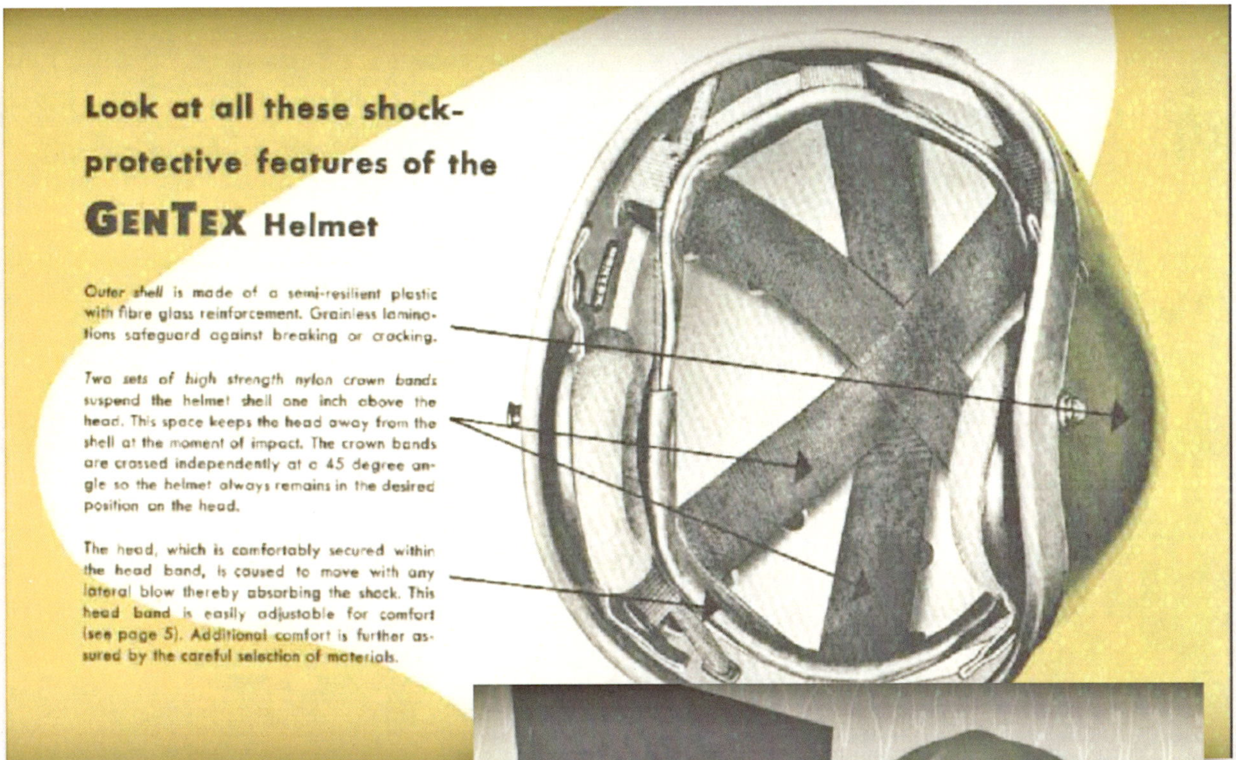

Without a shadow of a doubt, we have photographic proof that the Blue Angels used the GENTEX variants H-3 and H-4 in their support aircraft, but as of this book printing, no photos could be located to suggest the they were used in the demonstration aircraft.

As a retired 35-year career investigator, I am often reminded of this mantra from a colleague from the 1980s: "Absence of evidence, is not evidence of absence." These are profound words that remind us that there is often more to a story than meets the eye. We need to be driven by the thirst for knowledge. The passion for information, facts and the entire story should never be quenched.

To that end, the research must continue.

"Absence of evidence is not evidence of absence."

Unknown Author

6
USAF P-1 Flight Helmet

Following the development of the H-2 flight helmet, GENTEX introduced the H-3 and the H-4 variants (Chapter 5) which continued to feature the six radial reinforcing ribs. However, the helmets took on a radical design change featuring a two-piece approach, an outer hard shell coupled with a cloth inner liner very similar to the cloth helmets of WWII.

As discussed in chapter 5, these helmets saw limited use with the Team. There are unconfirmed reports that the Blue Angels felt the H-3 and H-4 helmets were not very comfortable so when their H-1 and H-2 predecessors were phased out, they chose a rather unconventional replacement. A P-1...from the inventory of the United States Air Force!

The P-1A helmet on the left is from the Smithsonian Air & Space Museum in Washington, DC., and donated by Fred Griffith.

The first contract for the P-1 series of flight helmets was awarded to the Switlik Parachute Company in 1948 with subsequent contracts to Lite Manufacturing and General Textile Mills (GENTEX).

The label (inset) is a typical example, but not directly associated with this particular helmet.

The P-1 — and variants, P-1A and P-1B — had served the U. S. Air Force well in the early 1950s, including the USAF Thunderbirds.[1]

The photo on the left is of LT "Lefty" Schwartz from

1957, who is holding a yellow P-1 helmet in classic Blue Angels markings of the period. This is one of only two vintage photos that has surfaced showing a P-1 helmet with a uniformed Blue Angel pilot. The inset photo — from the National Museum of Naval Aviation — is another example of this very unique helmet. (Author's photo)

The graphics on the front of the helmet feature the same style U. S. Navy wings that were used on the H-3 and H-4 helmets, and included the Blue Angels "banner" or "scroll" below the wings.

---

[1] *The Red Helmet - History, Development and Evolution of the USAF Thunderbirds Flight Helmets*, by Jeffrey D. Guidry

United States Navy Blue Angels Flight Helmets

During early March of 2020, I was able to arrange a meeting with the curator of the National Museum of Naval Aviation at NAS Pensacola. I discussed my research needs with the museum and they agreed to collect all of the Blue Angels flight helmets in archives and deliver them to the library for discussion and photographs.

I have lost count over the years of how many visits I have made to the National Museum of Naval Aviation, but trust me, it has been more times than any one person deserves! Each visit has offered me the predictable satisfaction that can only be delivered by one of the finest aviation museums on the planet while continuing to open my eyes and mind to aircraft, exhibits and artifacts discovered for the first time.

This visit would be different. To have a private research appointment with the curator of this world-class educational facility was not to be taken lightly.

Due to base restrictions[2], entry was limited to only those persons who were issued Department of Defense identification, something I did not possess. Not to worry, Dina G. Linn, Curator of the National Museum of Naval Aviation was permitted to sponsor my visit which allowed me access as a researcher.

I was particularly interested in the P-1 helmet on the previous page that I photographed several years ago during one of my many museum visits. Although no longer on display, I would soon be granted unrestricted access to this rare and intriguing artifact. I was woefully mistaken.

Sadly, this P-1 was not in archives, and there were no records to indicate who it belonged to, the identity of the donor or why it was no longer part of the museum collection. I was offered two possible scenarios: 1.) The helmet may have succumbed to previous hurricane damage sustained to the museum, or 2.) The owner or donor loaned it temporarily to the museum and has since retrieved the helmet. Whatever the case, another piece of history has vanished along with a story that yearns to be told. There's a part of me that wants to save them all!

---

[2] December 2019 active shooter incident; COVID-19 pandemic.

The only other photo I was able to locate of the P-1 flight helmet in use by the Blue Angels was the one on this page, from the John Slezak Collection. The pilot is Captain Charles E. "Chuck" Holloway, Jr., USMC. Capt. Holloway joined the Blue Angels in February 1956 and tragically lost his life during a practice demonstration in October of the same year.

What makes this specimen even more rare is the fact that the shell is painted white, not yellow.

Gone but, not forgotten.

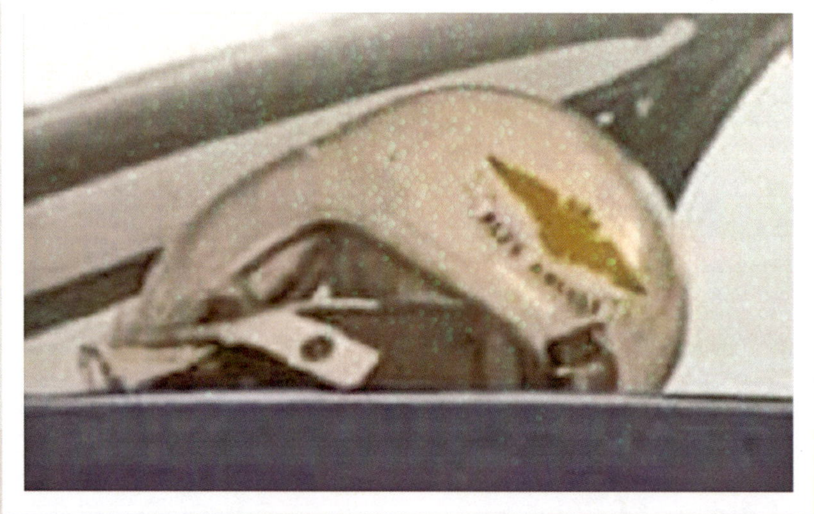

As a former law enforcement officer, detective and investigator with a major U.S.-based global energy company, I have amassed 35 years experience interviewing victims and witnesses, searching for clues and re-constructing a wide range of crimes and cases with little or no evidence or viable leads. I am driven by the thrill and challenge of sorting through and making sense of a jigsaw puzzle with missing pieces. This same drive also fuels my interest in collecting and preserving military aviation flight gear.

I mention this to help the reader understand that piecing together the history of the Blue Angels flight helmets is not a matter of surfing the Internet and stitching together the treasure trove of information and historical photos available at one's fingertips. Much of this history — a case-in-point is the Blue Angels' use of the USAF P-1 — is a challenge to gather, assemble and more importantly, unveil the whole story. More often than not, we are left with more questions than answers, as evidenced in this chapter. All too often, the easy work dries up quickly. This is what is referred to as "low hanging fruit." Unfortunately, this can lead to discouragement or boredom, and the case or research project ends up in the "cold case file" where it sits...sometimes indefinitely.

Much of this history is not written or recorded, rather it is filed away in the minds and memories of those pilots and support personnel who talked the talk and walked the walk. Regrettably, many of these eye witnesses have passed on, taking their precious knowledge with them, underscoring the importance of capturing and recording these living histories at every possible opportunity. It is becoming increasingly important and critical that this information is harvested and shared.

One such notable effort is by Ryan Nothhaft, a Blue Angels historian, videographer and grandson of Butch Voris, the first Blue Angels' "Boss." Ryan owns and manages a company called "Blue Angel Phantoms"[3] and spends a great deal of his time producing documentary videos on the Blue Angels, relying heavily on interviewing the pilots (and others) who have served with the Team over the years. These short films are brilliantly produced, awe inspiring and provide the viewer with an introspective look into the personal lives of the Blue Angels, often revealing historical accounts never before recorded.

[3] www.blueangelphantoms.com

Ryan's Web site is all about "Blue Angels history told by those that lived it." Ryan says, "Growing up, I had the great privilege of hearing the amazing origins of the Blue Angels Flight Demonstration Team from my grandfather, Butch Voris, who served as the team's first leader back in 1946. My appreciation and passion for the team's history has led me to initiate the Blue Angel Phantoms project, documenting the stories of Blue Angels alumni. It is my mission to share these stories as well as bring greater awareness to the excellence that is required to serve on the team."

"Blue Angels history told by those who lived it."

Ryan Nothhaft

You are to be commended, Ryan.

I have also enjoyed and benefitted from my own personal interviews with pilots and ground crew members while conducting research for this book. In addition to the passing of so many Team members who have served, many more are struggling with faded memories. Perhaps a sad commentary, but we need not be defeated or discouraged. We should do ourselves a favor by re-doubling our efforts and reaching out to spouses, children and grandchildren to capture what we can to preserve these fragile histories. As the search goes on, I pledge to continue my pursuit of knowledge in hopes of finding answers from those eager to share their stories.

It has been my personal experience that most of those people I interviewed for this and my companion book on the USAF Thunderbirds were very excited to tell their personal stories to an interested and willing listener. All too often, the information we seek is available...all we have to do is ask.

Just prior to this book going into production, I stumbled across a lead that the San Diego Air & Space Museum (San Diego, California) may have some Blue Angels flight

Photo courtesy of Becky Conrad, Assistant Registrar
San Diego Air & Space Museum, San Diego, California

helmets in their inventory. I made contact with Becky Conrad, Assistant Registrar, who graciously provided me with the beautiful photos on the following pages.

This is indeed an exceptionally rare and stunningly beautiful example of that elusive USAF P-1 flight helmet favored by the Blue Angels in the early to mid 1950s. This example belonged to "Zeke" Cormier, who served as the Blue Angels' "Boss" for the 1954, 1955 and 1956 show seasons. This helmet shows considerable wear, and the U.

S. Navy "wings of gold" positioned above the Blue Angels "banner" or "scroll" has faded with the passing of time. Upon closer examination, I determined that this P-1 is the same example that had been previously displayed at the National Museum

of Naval Aviation at NAS Pensacola, and was mysteriously no longer in their archives.

The rear and sides of this helmet can now be seen, studied and appreciated for her true beauty. Consistent with Blue Angels tradition, the rear of this helmet bears a name or call sign, in this case, the Boss' first name, "Zeke."

Thanks to the San Diego Air & Space Museum for not only preserving these very rare artifacts but, making a concerted effort to share them with the general public.

7
APH-5 Flight Helmet

A June 1956 article from *Naval Aviation News* reported "After more than two years of planning and testing, BuAer (Bureau of Aeronautics) announces the development of a new protective helmet, the APH-5. It is so designed as to insure comfort and close fit at the same time it provides the protection and functionality required for high speed flight."

"Scientific progress now permits the incorporation into the laminated fiberglass helmet of a rigid plastic foam material which yields at a given force. In crushing, it dissipates the force and protects the head. It offers far greater protection than the old suspension-type helmet."[1]

[1] Referring to the GENTEX H-3 and H-4 variants.

"It is cooler and more comfortable than its predecessors. Foam liners come with each headpiece so that there will be little or no difficulty in fitting the pilot. Each helmet is equipped with a retractable eye shield assembly which will accommodate either a clear or neutral gray filter lens. This was designed for maximum vision and wind-blast protection."

"The helmet's ear cup assembly reduces noise to a minimum. The chin strap and oxygen mask mounting tabs have been designed to retain the helmet and mask during emergency escapes at high speeds."

"VX-3 at NAS Atlantic City, the Aero Medical Branch of Service Test Division at NATC Patuxent, Air Crew Equipment Laboratory, Naval Air Materiel Center, Philadelphia, and AirLant and AirPac Fleet units were among those who cooperated with the Airborne Equipment Division of BuAer in testing the gear."

"The distribution of the helmets began in April, the first going to fleet squadrons as allocated by ComAirLant and ComAirPac. BuAer has announced that eventually they will become standard equipment in all fighter and attack squadrons."

According to the NAVPERS (Navy Personnel Command) 10358-A, Parachute Rigger 3 & 2 NAVY Training Course, the APH-5 is the U. S. Navy's newest protective flight helmet. The helmet is available in large and medium sizes. The complete unit consists of an outer shell, energy-attenuating liner, sizing pads, integrated visor, ear cup assemblies, communication cord, oxygen mask mounting tabs, nape strap, and adjustable chinstrap.

The outer shell provides force-distribution and penetration-resisting properties, and is molded from glass fabric and a polyester resin. The energy-attenuating liners are provided to absorb or dissipate impact energy and thus prevent skull fracture or brain injury, or both. The three sections of the liner, which are made of cellular polystyrene sheets, are molded to fit closely the inside contours of the outer shell.

The sizing pads permit fitting of the helmet to an individual's head contour with the proper balance between stability and comfort. The sizing pads are supplied in the basic configurations – back, crown, and front, also in two basic thicknesses – thin and thick. The pads are leather covered, miniature size to improve comfort by reducing the discomfort that could possibly result from internal heat in the helmet. Aviation Clothing and Survival Equipment Bulletin No.3-57A, authorizing modification of the sizing pads, must be complied with on all helmets that are in service and not already modified.

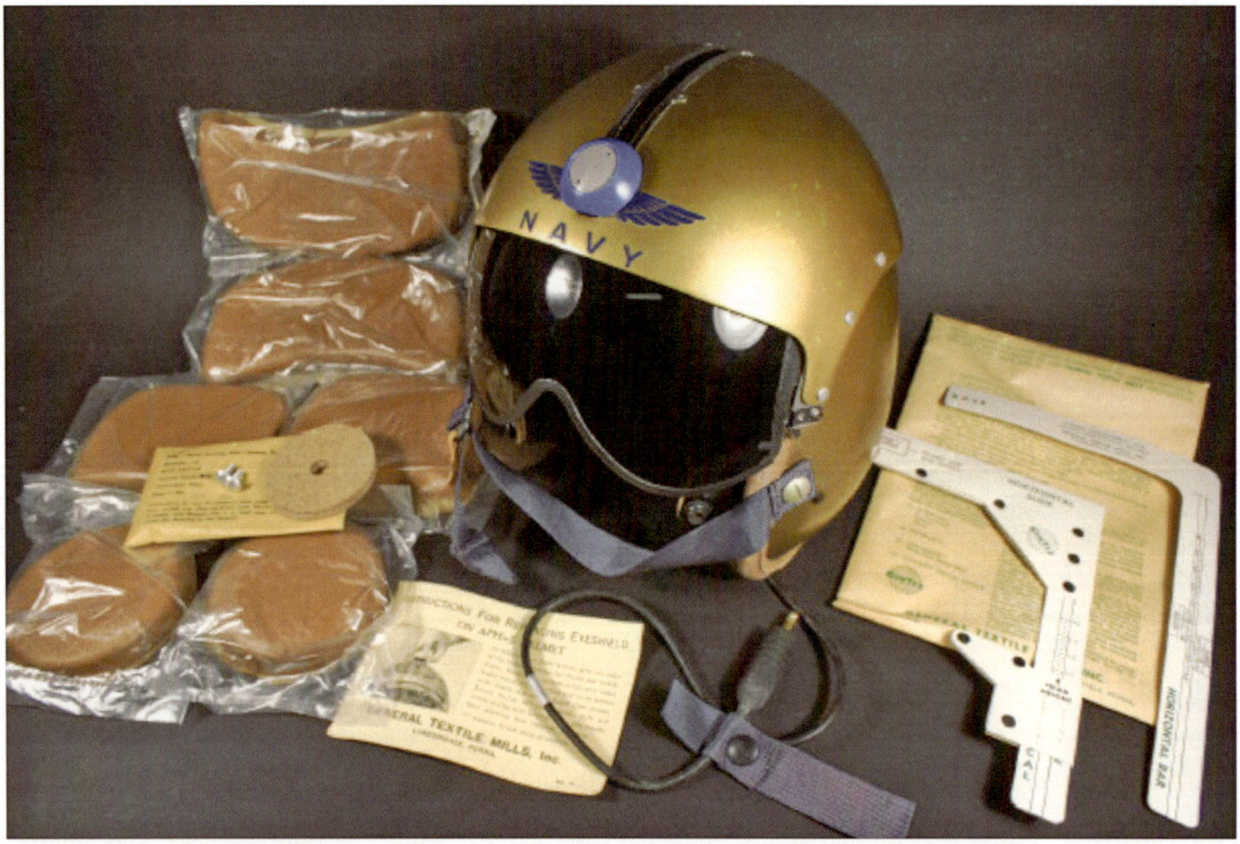

(Photo credit: http://www.best-of-flightgear.dk/

The integrated visor provides protection from sun glare, wind blast, dust, foreign particles, and flash fires. It has excellent optical properties and crack propagation resistance and is protected from damage when in the UP position by a housing. The housing is molded from a glass fabric and a polyester resin, similar to the outer shell.

Two types of visors are available for use in the helmet assembly; clear and neutral grey. The neutral grey visor, which is darker at the top than at the bottom to afford maximum protection from sun glare and to facilitate reading of the instrument panel, is currently supplied in various densities. This range of densities satisfies various requirements for special visual conditions and / or individual preferences dictated by such factors as the demand by blond or light pigmented individuals for darker lenses. The relative darkness or degree of visible light transmittance is indicated by white markings on the inside edge of the visor. This marking can be seen more readily by removing the visor from the housing.

Due to the manufacturing processes, it is impractical to control transmission of the various filter visors produced. In view of this practice, medium and large size visors with varying percentages of light transmittance are carried in the supply system under a single stock number. It is recommended that personnel check lenses at squadron level to obtain individual preferences before obtaining from supply.

The manufacturer's tags below - Mine Safety Appliances Company, and General Textile Mills, Inc., both had contracts to supply the U. S. Navy with APH-5 protective flight helmets.

On the following three pages are early patent drawings of the APH-5 visor housing and adjustable visor mechanism.

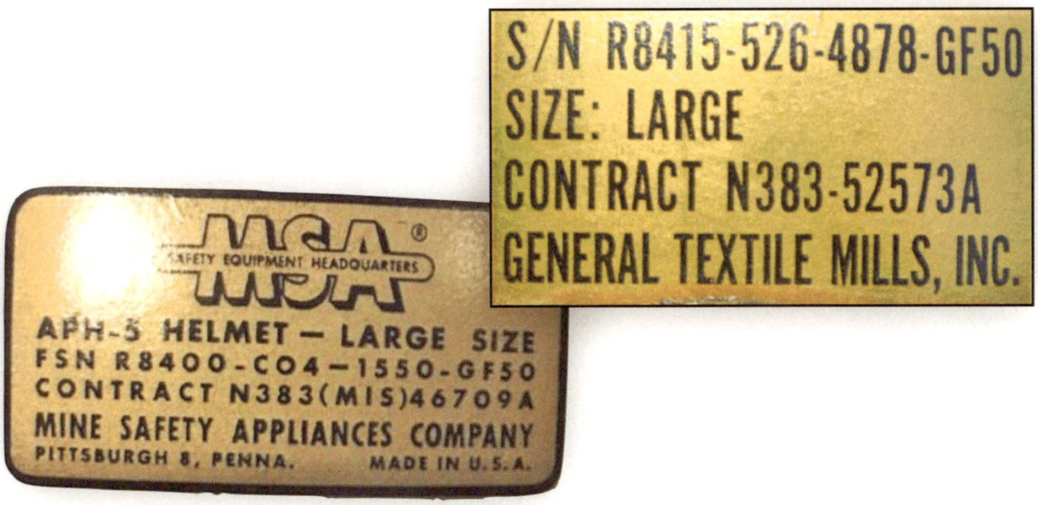

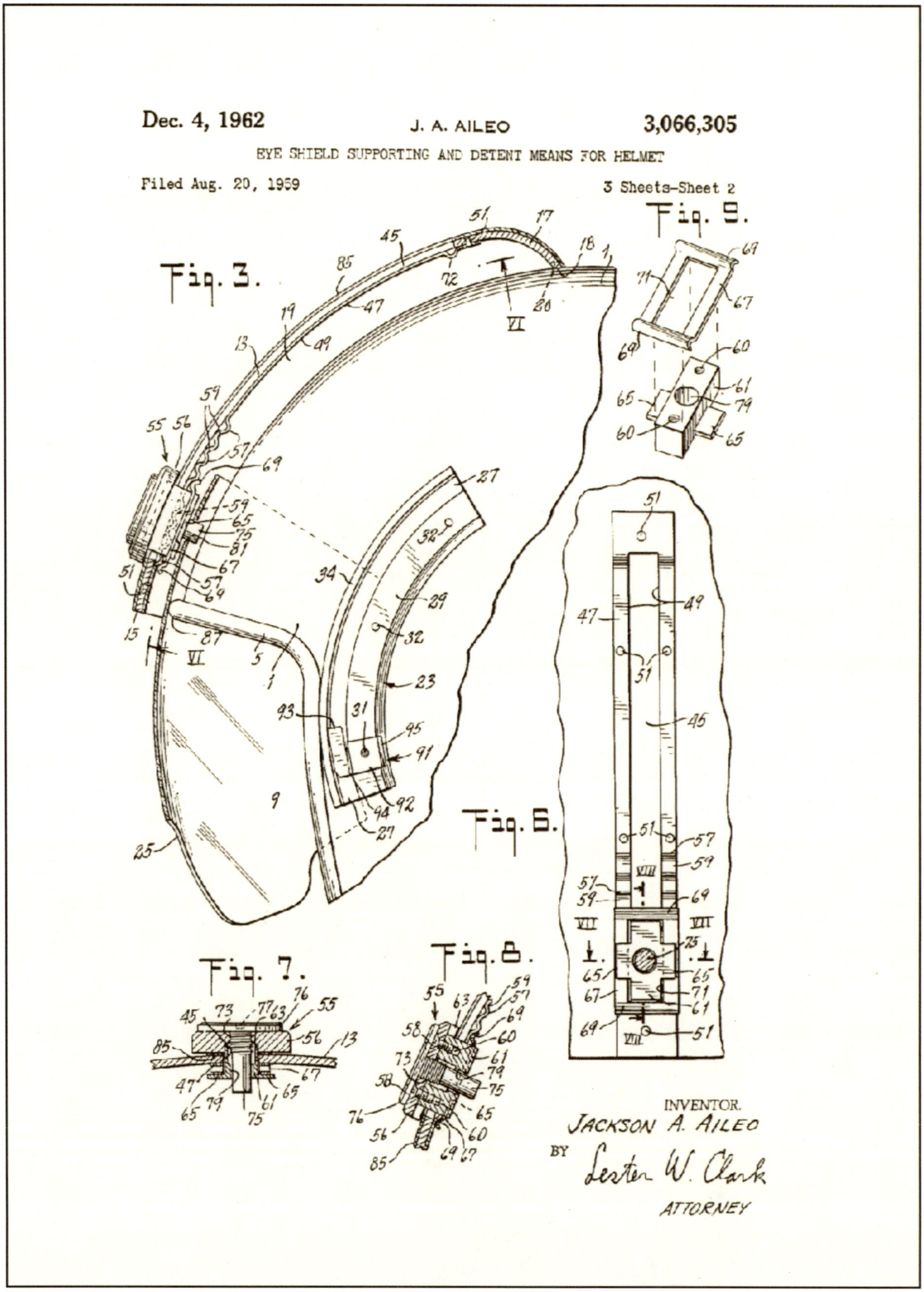
Dec. 4, 1962
J. A. AILEO
3,066,305
EYE SHIELD SUPPORTING AND DETENT MEANS FOR HELMET
Filed Aug. 20, 1959
3 Sheets-Sheet 2
Fig. 3.
Fig. 9.
Fig. 6.
Fig. 7.
Fig. 8.
INVENTOR.
JACKSON A. AILEO
BY Lester W. Clark
ATTORNEY

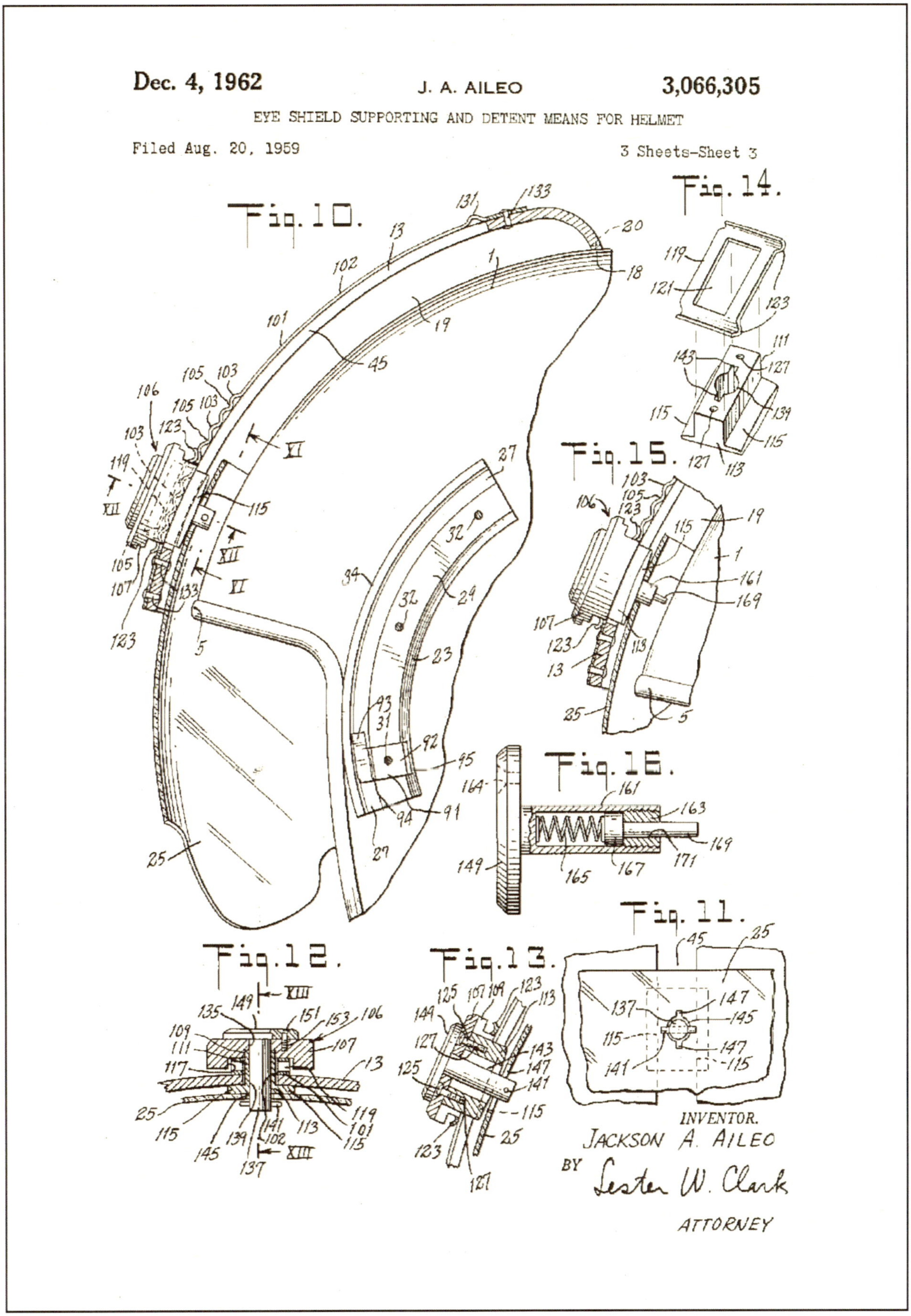
Dec. 4, 1962
J. A. AILEO
3,066,305
EYE SHIELD SUPPORTING AND DETENT MEANS FOR HELMET
Filed Aug. 20, 1959
3 Sheets-Sheet 3
Fig. 10.
Fig. 14.
Fig. 15.
Fig. 16.
Fig. 11.
Fig. 12.
Fig. 13.
INVENTOR.
JACKSON A. AILEO
BY Lester W. Clark
ATTORNEY

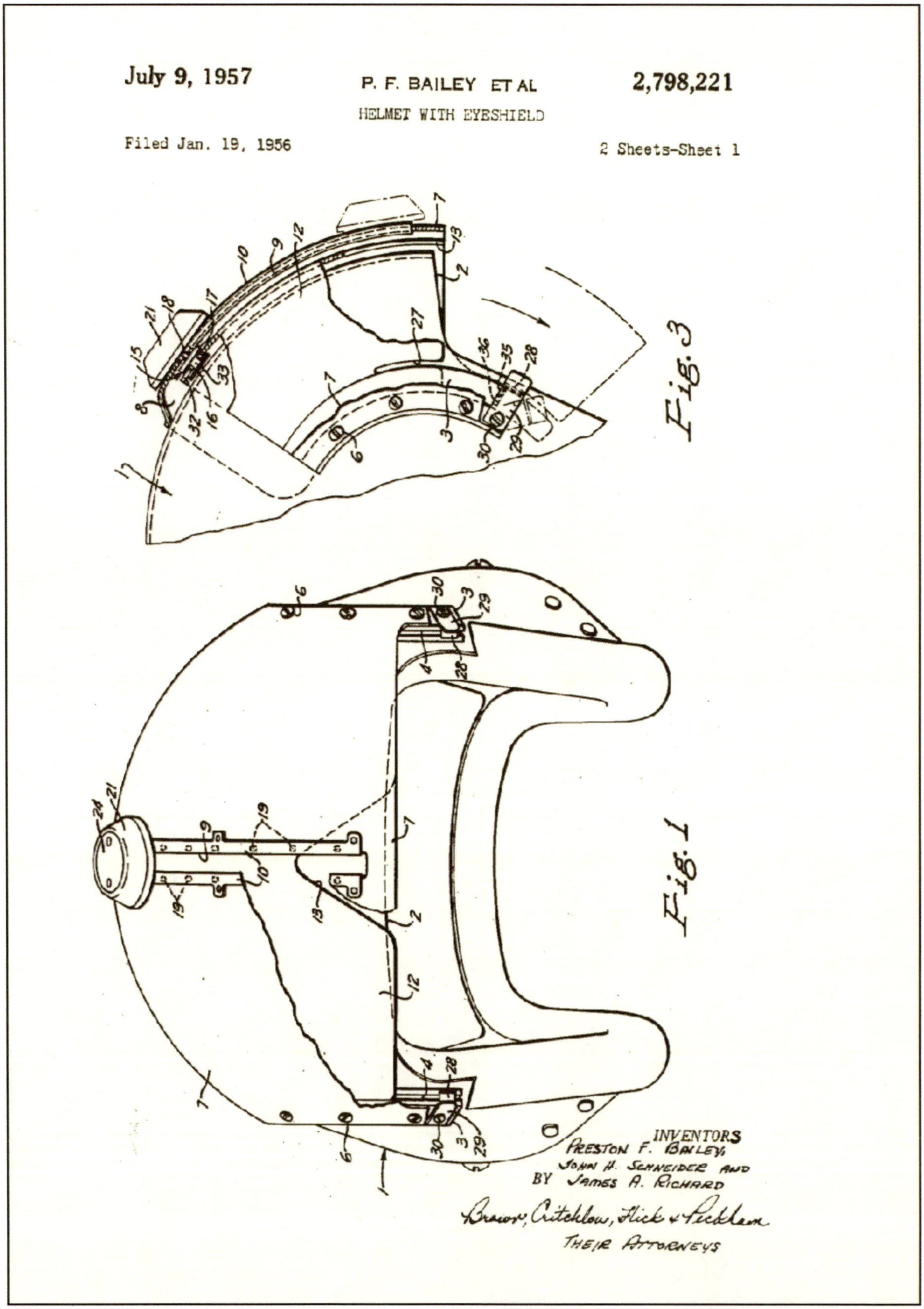
July 9, 1957
P. F. BAILEY ET AL
2,798,221
HELMET WITH EYESHIELD
Filed Jan. 19, 1956
2 Sheets-Sheet 1
Fig. 3
Fig. 1
INVENTORS
PRESTON F. BAILEY,
JOHN H. SCHNEIDER AND
BY JAMES A. RICHARD
Brown, Critchlow, Flick & Peckham
THEIR ATTORNEYS

U. S. Navy Blue Angels APH-5 Flight Helmet

U. S. Navy Blue Angels, 1957 (U. S. Navy photo).
Team members included LT Herb Hunter, LT Nello Pierozzi, LT Sheldon Schwartz, LT (jg) Bob Rasmussen, LT Thomas Jefferson, and CDR Ed Holley (team leader - far right)

By early 1957, the APH-5 flight helmet began to see widespread distribution to U. S. Navy flight crews throughout the fleet. The two-piece outer hardshell / soft inner liner setup unique to the H-3 and H-4 helmets gave way to a hard shell helmet with inner sizing pads. This must have met favorably with the Blue Angels, because they abandoned their USAF P-1s and welcomed the new U. S. Navy APH-5.

The standard issue APH-5 helmets were typically painted gold, as seen on the first page of this chapter, and included "split wing" and "NAVY" decals on the visor housing. The Blue Angels painted their gold helmets yellow and used the "split wing" decal but, opted for the Blue Angels "scroll" instead of the "NAVY" decal, as seen in this photo (restored APH-5 in author's private collection).

The early APH-5 helmets offered some innovative features including an external visor with a fiberglass protective housing. A push-button mechanism allowed the wearer to raise or lower the visor and lock it in any number of positions along the chrome plated edged center track, a unique feature to this helmet.

The lower portion of the visor was covered with a rubber edge bead, and the slotted sides rode along a steel guide to help stabilize the visor in the full down position (center and bottom photos). The visor housing was attached to the helmet shell with 3 screws and studs with metal visor tracks "sandwiched" in between that later gave way to nylon and plastic tracks.

These early APH-5s also used the leather oxygen mask connection tabs with snaps which was the standard at the time.

Later versions would eliminate the leather tabs and incorporate nickel plated oxygen mask bayonet receivers manufactured by the Hardman Tool & Engineering Company. Ratcheted "T" shaped bayonets — often referred to as "Christmas Tree bayonets — and webbing assembly served a dual retention purpose: to secure the oxygen mask to the helmet, and to secure the helmet to the wearer's head (before the addition of a chin strap).

United States Navy Blue Angels Flight Helmets

In April of 1962, the U. S. Army Transportation Research Command, conducted a "Crash Injury Evaluation" study on the APH-5 Helmet. According to the report, the APH-5 is manufactured in accordance with Navy specifications. The development of an improved protective helmet for air crews was initiated by the United States Navy through several independent research projects.

U. S.Navy photo (Author's private collection) of the Blue Angels Team with their APH-5 flight helmets and a Grumman F11F "Tiger" dated 1961.

The report findings identified no less than 20 areas of deficiency, including weak geometric strength, insufficient energy absorbing liner qualities, weak retention system and overall discomfort to the wearer due to lack of ventilation and heat dissipation, just to name a few.

By the time this U. S. Army report was published, the U. S. Navy had already begun developing a replacement helmet, designated APH-6 (Chapter 8).

The photo above is a *Union Bulletin* (U-B) photo from 1960 of Blue Angels pilot LT Duke Ventimiglia (right) and U-B reporter, Jim Schnick holding an APH-5 flight helmet prior to an incentive ride.

LT Duke Ventimiglia, Blue Angels, at NAS Pensacola. LT Ventimiglia is perhaps inspiring this unidentified boy who is wearing a Blue Angels APH-5 flight helmet, as the next generation of naval aviators. (U. S. Navy photo).

In this undated photo, a young woman tucks her hair into an APH-5 flight helmet as she prepares for a VIP incentive ride. Note the visor housing bears the Blue Angels scroll but not the wings. (U. S. Navy photo)

So popular were the Blue Angels, that California National Productions produced a television series about the Team that aired on NBC[2] in 1960 and 1961. In these two promotional photos (from author's personal collection), actor Warner Jones, USMC, who plays Captain Bill Scott, poses for the camera in his Blue Angels flight suit and APH-5 flight helmet.

The back of the photo reads, "Capt. Bill Scott is the only Marine Corps flyer in the Navy foursome whose thrilling adventures are dramatized in the 'Blue Angels' television series. Scott is played by Warner Jones, a legitimate theatre actor who recently made his TV debut on 'Playhouse 90.'"

An NBC affiliate advertised the Thursday program in a local newspaper advertisement that read, "Knife through the skies with the U. S. Navy's celebrated four-man precision-flying team and share their stirring adventures on land in dramas brimming with action, suspense and human interest."

---

[2] NBC Television Films is a Division of California National Productions, Inc.

The Yellow Helmet

The Blue Angels appear to have used a couple of different wing designs on their APH-5 flight helmets during their use with the Team. One example was the standard issue U. S. Navy wing decals that were common to all APH-5 flight helmets of the period. The decal card included either NAVY or U.S.M.C. below the wings. The Blue Angels only used the wings.

The second example were the standard issue U. S. Navy wings commonly used on the H-3 and H-4 flight helmets. The wings were cut in half (vertically) and applied to either side of the chromed visor housing lock track. The photo below is a rare and beautiful example of this type (courtesy of the John Slezak collection).

APH-5

Table 14-1 – Visor marking and percentage of light transmittance.

Marking	Percentage of light transmittance
D	10 to 20 percent inclusive
XD1	8 to 10 percent (medium dark)
XXD	6 to 8 percent (dark)
XXX	5 to 6 percent (very dark)
*D	10 to 20 percent inclusive
*XD	Under 10 percent

* In the initial quantity of helmets issued to the fleet, these two types of neutral grey visors were installed.

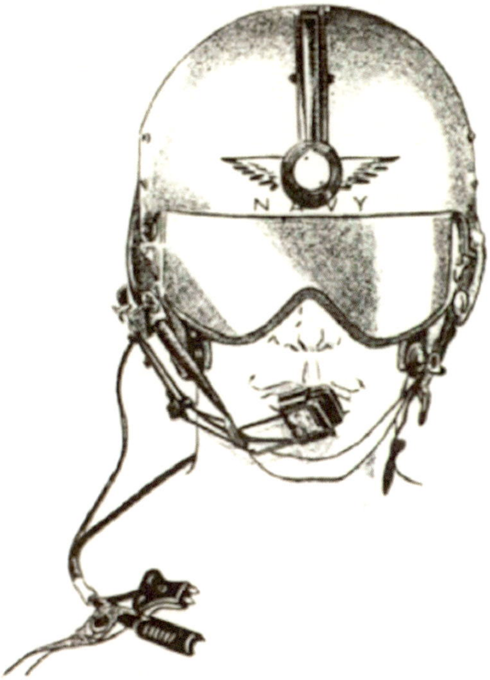

Figure 14-5.—Type APH-5 pilot's protective helmet.

This is a vintage chart describing the various visors available for the APH-5 flight helmet, listed according to the percentage of light transmittance. This pre-dates the gold, nuclear, anti-flash visors popularized by the Blue Angels, that were introduced in the early 1970s.

Just when you think all leads have been exhausted in what seems like a never ending search for information, you catch a break. While searching for examples of APH-5 flight helmets used by the Blue Angels on the Internet, I came across an example that the photographer tagged, "San Diego Air & Space Museum." This is what is known as a clue. I reached out to the San Diego Air & Space Museum inquiring about this helmet and received a reply

from Becky Conrad, Assistant Registrar, who graciously provided me with the photographs on this page. According to Ms. Conrad, the inside of the helmet is marked with the name, "Dave." The only pilot named "Dave" during the period the APH-5 was in use was Lt. Dave Rottgering, #7 in 1966 and 1967.

United States Navy Blue Angels Flight Helmets

A few final thoughts about the APH-5 flight helmet.

The three photos on this page illustrate the graphic design changes that took place on the visor housings of the APH-5. While difficult to date these variations, it is still interesting to recognize that there was a concerted effort to find the right "look" for these yellow canvases.

The top photo simply bears the Blue Angels "scroll" or "banner" and has stood the test of time, as this is the same design used on contemporary helmets.

The center photo includes the addition of U. S. Navy wings, a decal that was part of most U. S. Navy issued H-3 and H-4 flight helmets (See Chapter 5).

The bottom photo is a variation on a theme, using a more contemporary version of wings that were used on most U. S. Navy issued flight helmets of the period.

The image above is a U. S. Navy photo (author's private collection), dated May 1962. LT Hank Giedzinski, the narrator (#7) for the 1961 and 1962 show seasons is providing some last minute, in-cockpit instructions on the merits of the APH-5 flight helmet. The lucky passenger is Dick Vonier, an unspecified news reporter who is preparing to experience a Bluer Angels VIP ride in the back seat of an F9F-8 "Cougar."

8
APH-6 Flight Helmet

The Aircrew Protective Helmet, Version 6 — or APH-6 — was introduce between 1960 and 1961 and offered the wearer some improvements over its predecessor the APH-5. While APH-5 and APH-6 appeared very similar, the APH-6 is most noted for the "U" shaped, ratcheted oxygen mask retainer tracks, commonly referred to as "horseshoe" oxygen mask receivers. The oxygen mask was equipped with retainer releases that are known among collectors as "butterfly" clips because of their "winged" appearance (photo below right). The APH-6 was also lighter weight than the APH-5.

The APH-6 also opted for a thin, black rubber edge bead rather than the neoprene edge roll common to the APH-5. The early fiberglass visor and metal visor housing tracks, borrowed from the APH-5, were later upgraded with a plastic visor housing and nylon visor tracks.

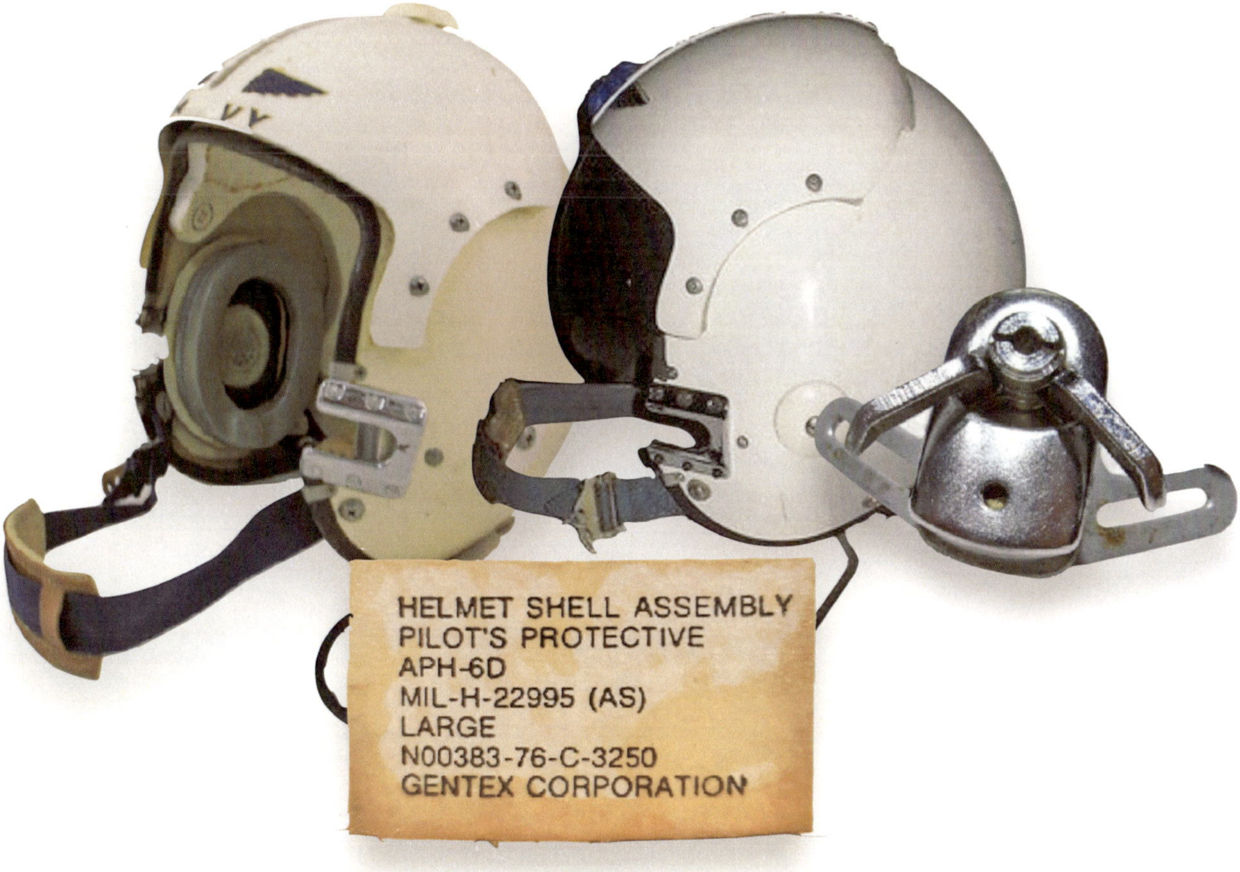

According to the Naval Training Command manual, NAVTRA 10358-E, Aircrew Survival Equipmentman 3 & 2, the APH-6 series of flight helmets were configured according to their application. The APH-6D helmet (single visor) is designated as the primary flight helmet for use in all aircraft except helicopters.

The APH-6D helmet provides eye protection, sound attenuation and protection for the wearer's head during in-flight buffeting, seat ejection, bailout or crash landings. It is designed to distribute impact forces over the entire helmet, and to absorb these forces so that a minimum amount of impact reaches the wearer's head.

The helmet consists of an outer shell assembly, clear or neutral visor assembly, as required and a sonic ear cup assembly. The outer shell assembly is molded from fiberglass and polyester resin and provides impact and penetration protection. The edgeroll is made of rubber and provides protection from the helmet edges. The outer shell assembly also includes the inner foam liner, sizing liners, communications cord set, oxygen mask retaining tracks, nape strap, and adjustable chin strap.

The APH-6, and later variants APH-6A (single visor), APH-6B (dual visor), APH-6C (single visor) and APH-6D (dual visor), referred to minor changes in earphones, ear cups and styrene liners. In later years, the nomenclature was modified to include the size of the helmet, e.g., APH-6DM and APH-6DL, where the "M" represented a medium helmet and the "L" represented a large helmet.

The APH-6 series of flight helmets served the U. S. Navy well for a couple of decades including through the Vietnam War. In a letter dated August 1978 the Chief of Naval Operations concluded the APH-6 "no longer meets operational requirements" but was not officially "cancelled without replacement" until August of 1992.[1]

Two major suppliers were granted contracts to furnish APH-6 flight helmets to the U. S. Navy, USMC and U. S. Coast Guard: GENTEX Corporation of Carbondale, Pennsylvania and Sierra Engineering Company of Sierra Madre, California.

In a 1968 GENTEX sales brochure,[2] the APH-6 helmet was described as being constructed from polyester impregnated fiberglass cloth laminate finished with a white epoxy paint. The vented visor housing is an injection molded, high impact thermoplastic. Although the retainer track for the oxygen mask retention and release assembly was standard, ***modified retention systems were available upon request.*** (Author's emphasis).

[1] best-of-flightgear.dk

[2] From the private collection of Steve Norris.

The APH-6 flight helmet photos on this page belonged to Dick Schram, who was with the Blue Angels from 1969 to 1971. U. S. Navy fleet pilots used the same type of flight helmet, but theirs did not include the same custom features afforded Blue Angel pilots, such as a hand-sewn leather edge roll, fitted leather liner and leather covered ear pads.

According to Robert Schwaiger — Blue Angels PR — Protection, Inc. began coming out to El Centro during winter training to custom fit helmets to the pilots beginning in 1967 or 1968.

Of course, the paint job was also very special and unique to these Blue Angel helmets.

The custom paint included an extra feature that was a signature of Protection, Inc. — silver metal flake embedded in the clear coat.

In the 1960s, visor housings from GENTEX and Sierra Engineering Company came standard with a straight, forward edge (see photo top left). Over time pilots began to report difficulty when looking high left and high right. Their view was being obstructed by the design of the visor housing.

To remedy this problem, the U. S. Navy developed a template (drawing below) in May of 1970 to "field modify" the visor housings by relieving the left and right temple areas, which resulted in improved vision. This "gull wing" modification was a success and flight helmet manufacturers began

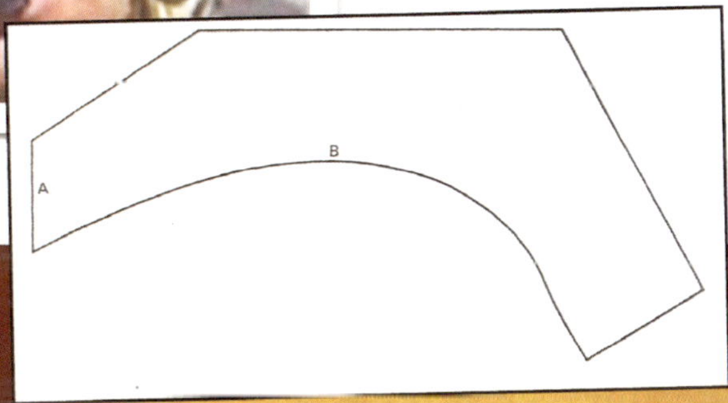

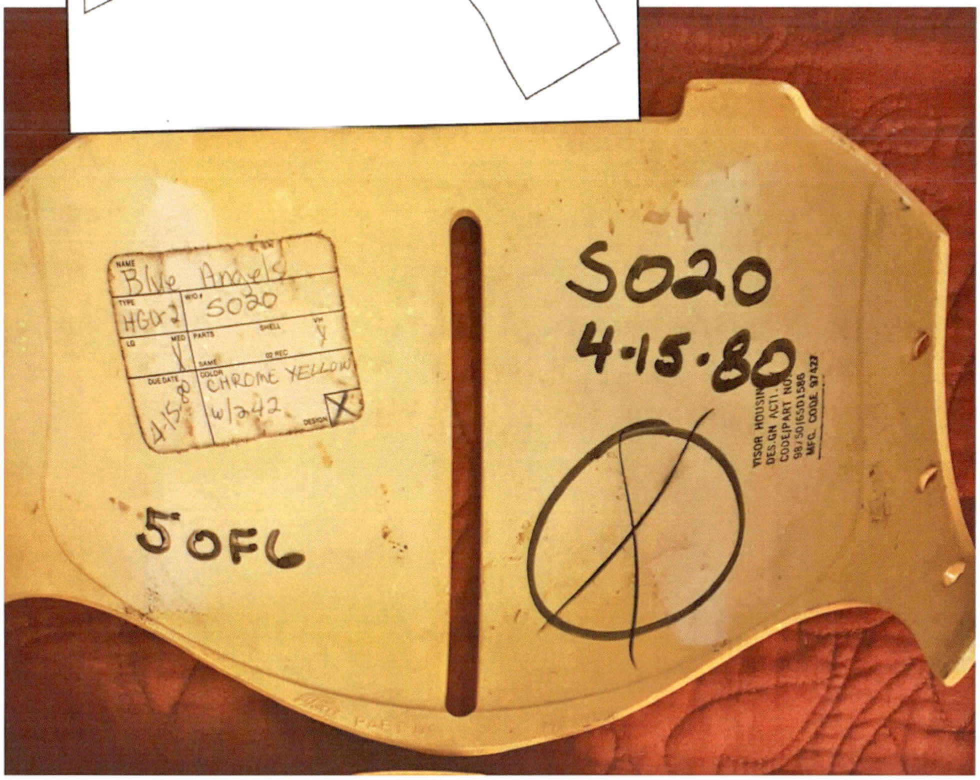

producing their visor housings with this modification from the factory in the mid to late 1970s.

The photos on this page help to illustrate the visor housing modifications prior to becoming a factory redesign.

The top left photo is a standard issue APH-6 helmet visor with an unmodified, straight front edge.

The second photo is an APH-6 helmet visor that has masking tape applied and marked using the field modified visor housing template.

The third photo provides a visor housing field modification in progress.

The fourth photo is an APH-6 visor housing upon completion of the field modifications.

All photos and field modifications on this page are courtesy of and completed by AW1(NAC/AW) Michael H. Johnson, Jr., U. S. Navy (Retired).

Pictured above is LT Hal Loney, Blue Angel solo pilot in 1968. This is a contemporary photo of Loney wearing his vintage "Blue Suit" and posing proudly on his Grumman F-11-F-1 "Tiger" show aircraft. On the canopy rail is his APH-6 flight helmet. The aircraft, artifacts and photo are from the Combat Air Museum in Topeka, Kansas.

I reached out to Kevin Drewelow, Director of the Combat Air Museum requesting assistance with obtaining some high resolution photos of LT Hal Loney's "blue suit" and APH-6 flight helmet. He pressed Klio Hobbs into service, who provided all of the beautiful photos on the following two pages.

Photos courtesy of the Combat Air Museum, Topeka, Kansas.

Photos courtesy of the Combat Air Museum, Topeka, Kansas.

Over the years, the APH-6 underwent some "unofficial" field modifications and improvements namely with the oxygen mask retention mechanism. The original APH-6 helmets were equipped with a very unique U-shaped retainer track, often referred to in collector circles as a "horseshoe" oxygen mask receiver. This assembly was paired with a retainer release mechanism, attached to the oxygen mask straps, that was commonly described as "butterfly" clips (pictured left). These were eventually outdated and abandoned.

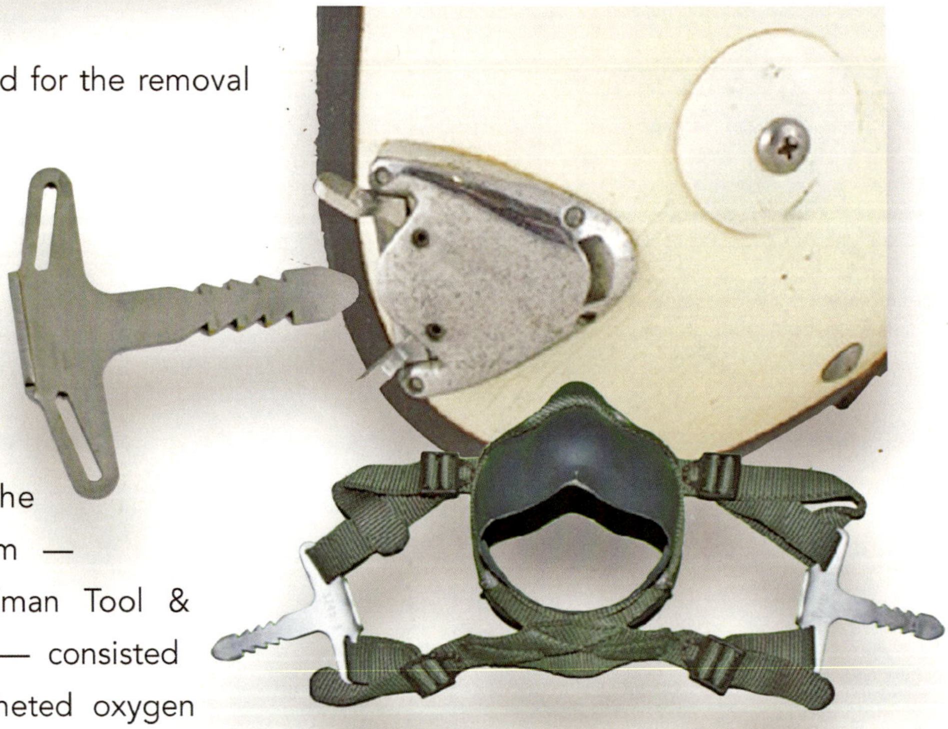

Field modifications called for the removal of the "horseshoe" tracks, and the void to be filled with automotive body filler or similar product. The patch was sanded, primed and painted to match the helmet. The replacement mechanism — manufactured by Hardman Tool & Engineering Company — consisted of a "T" shaped, ratcheted oxygen mask bayonet that secured the oxygen mask (via webbing) to a nickel plated bayonet receiver mechanism attached to the helmet shell (pictured right). Only a few examples of this field modification have showed up in collector circles.

The field modifications to the helmet shell were to allow for the upgrading to Hardman oxygen mask receivers. As it turns out, the APH-6 — manufactured by GENTEX and Sierra Engineering, Company — was offered with the "horseshoe" oxygen mask retainer cutouts, or without, to allow for more contemporary retainers.

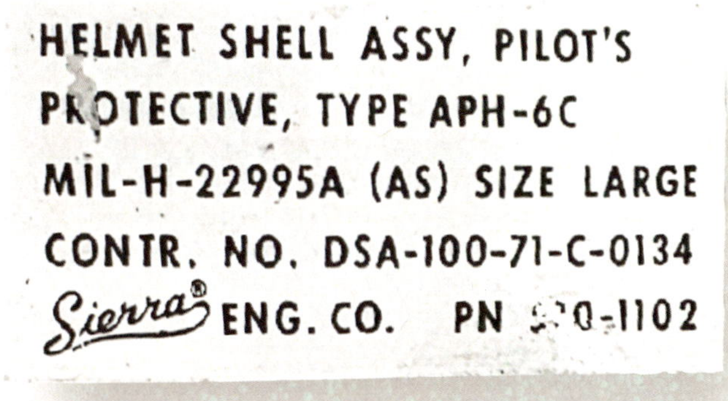

In December of 1976, the Department of the Navy, Naval Air Development Center (NADC)[3] filed a report on the findings from a study titled "AIRCREW PROTECTIVE CLOTHING & DEVICES SYSTEM (FIGHTER / ATTACK AIRCRAFT)."

The report abstract read as follows:

> **A study was conducted to portray the major mission profiles in which the Fighter / Attack community is engaged and by so doing, to identify the problem areas associated with the current inventory of Fighter / Attack Aircrew Protective Equipment toward fulfilling the requirements of these missions. The ultimate objective of this effort is to provide a sound basis from which a series of separate but coordinated engineering developments will be conducted to provide a new generation of mission-oriented protective systems for the Fighter / Attack community. The net effect of this program will be to enhance the inflight performance and effectiveness of all Fighter / Attack aircrewmen with little or no sacrifice to their safety in the event of an emergency situation.**

3 Report No. NADC-75294-40

The study covered a comprehensive list of flight crew protective equipment including the APH-6 flight helmet. Through extensive observations, discussions and data reviews, the report cited a broad range of conflicting data and differences of opinion concerning specific equipment problems, resulting in a number of adverse effects on aircrew performance and overall mission effectiveness. The study also found that these problems were compounded when the crew member made unauthorized, but sometimes necessary modifications to equipment configuration to "self-enhance" the equipment effectiveness.

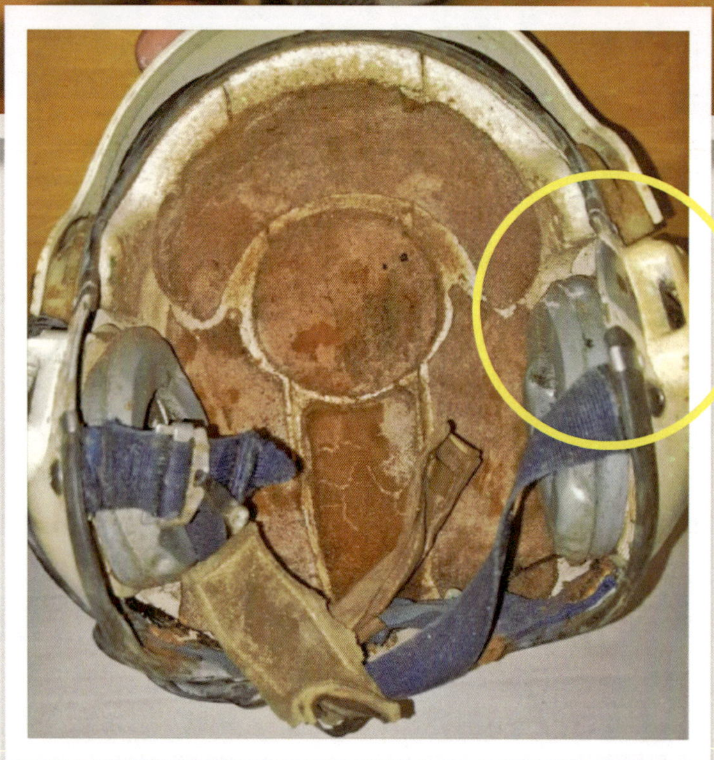

The two photos on the left (from a 2020 eBay listing) show an example of an APH-6 helmet with the "horseshoe" cutouts filled to accept a more contemporary cast aluminum oxygen mask receiver.

Regarding the APH-6, the flight helmet was described as having five basic functions, which included (in part) to providing head protection from forces of impact, buffeting and vibration; Provide protection to the face and eyes by way of a Visor System; Provide hearing protection; Provide a convenient base for the attachment of oxygen mask and communication systems; and to provide protection against cold exposure in a survival situation.

The report identified several specific problem areas. The most frequent complaint from crew members concerned sizing, fit and comfort. Despite the availability of a wide selection of leather-faced foam rubber sizing pads, and adjustable chin and nape straps, in many cases a satisfactory fit could not be achieved and / or maintained for sustained periods of time. The weight of the helmet — approximately 4.5 pounds — exacerbated the problem, often leading to head and neck fatigue. Additional problems included excessive heat buildup, resulting in increased perspiration, contributing to premature fatigue and dehydration.

Other reported problems included visors that were too dark, highly vulnerable to chipping and scratching, and clear visors that distorted the wearer's visual field.

The report recommended the pursuit and testing of new and emerging lightweight materials for helmet shell candidates, citing many contemporary materials offer higher strength-to-weight ratios, and are lighter and stronger than the current APH-6 fiberglass shell.

Further recommendations suggested an emphasis on form-fit liners rather than pad set liners, further ensuring helmet comfort, fit and stability for sustained periods of time.

The study concluded that these recommendations are offered with the realization that the design and development of a helmet is a difficult and complex effort which can only be conducted on an incremental trial and test basis.

Fortunately, for the Blue Angels, their flight helmets were being customized by Protection, Incorporated (PI), a subsidiary of GENTEX. Because the Blue Angels were a relatively small squadron and funded a little differently from their sister squadrons in the fleet, they were afforded cutting edge upgrades to their helmets that were not standard U. S. Navy-issue.

The Team enjoyed the benefits of professional paint jobs, custom hand-sewn leather edge rolls and custom, form-fitted leather liners, considered a luxury for the time period.

For the fleet pilots, the complaints continued. In a June 1977 article from *Approach Magazine*[4] titled, "The Pilot's Dilemma," LCDR Scott Stallings[5], of VF-124 reiterates the cries of the fleet pilots about their flight gear. He described the process of flight gear issue and fitting as "haphazard at best" and "the combination of limited equipment size and lack of customization make optimum fitting difficult."

He described flight helmets — in particular, the APH-6 — as a "necessary evil" further citing that "They are hot, bulky and generally uncomfortable." The APH-6 was also characterized as "heavy" which requires additional strain on neck and back muscles at high-g loading.

LCDR Stallings suggested that every crew member should be issued a form fit flight helmet such as those customized by Protection, Inc.[6] He added, "There are no small, medium or large heads, and the cost of a form fit helmet compared with its value in terms of comfort, safety and tactical advantage make it well worth the investment."[7] LCDR Stallings' U. S. Navy-issued APH-6 flight helmet was outfitted with a form fit liner and he cherished it so much, he took out an insurance policy on it in the event it was lost or damaged, while other pilots were known to pay for upgrades out of their own pockets.

LCDR Stallings said, "I am a pilot, not an engineer" suggesting he was not aware of all the challenges associated with flight gear design and development, and doubted the engineers were aware of of the problems associated with wearing the flight gear they

---

[4] Provided by Steve Norris.

[5] Killed 15 June 1978 in an F-14A "Tomcat" crash at sea near San Clemente Island, California.

[6] Chapter 18, "Protection, Incorporated"

[7] Form Fit helmet modifications from Protection, Incorporated were reportedly $250 to $300 per copy.

design. LCDR Stallings suggested, "Instead of throwing rocks at each other, let's open up a dialogue to achieve our common goal."

The photo on the right is a final example of an APH-6 that supports GENTEX's sales brochure claim that these helmets were available with modified retention systems upon request. Similar to the APH-6s with tracked cutouts, the shell is equipped with a thin black rubber edge bead with neoprene padding at the nape. This example also retains the "adjustable" ear cup feature popularized by the APH-6 family of flight helmets. The screw hole through the helmet shell was approximately 3/4 inch in diameter to allow for the ear cups to be adjusted to the individual wearer. The screw was fitted with a plastic disk to cove the hole.

In Chapter 9 we take a look at work that GENTEX had undertaken a few years prior to this study, and their response to address similar problems stemming from the end of the Vietnam War.

(Ear cup and ear cup disk photo credit: flighthelmet.com).

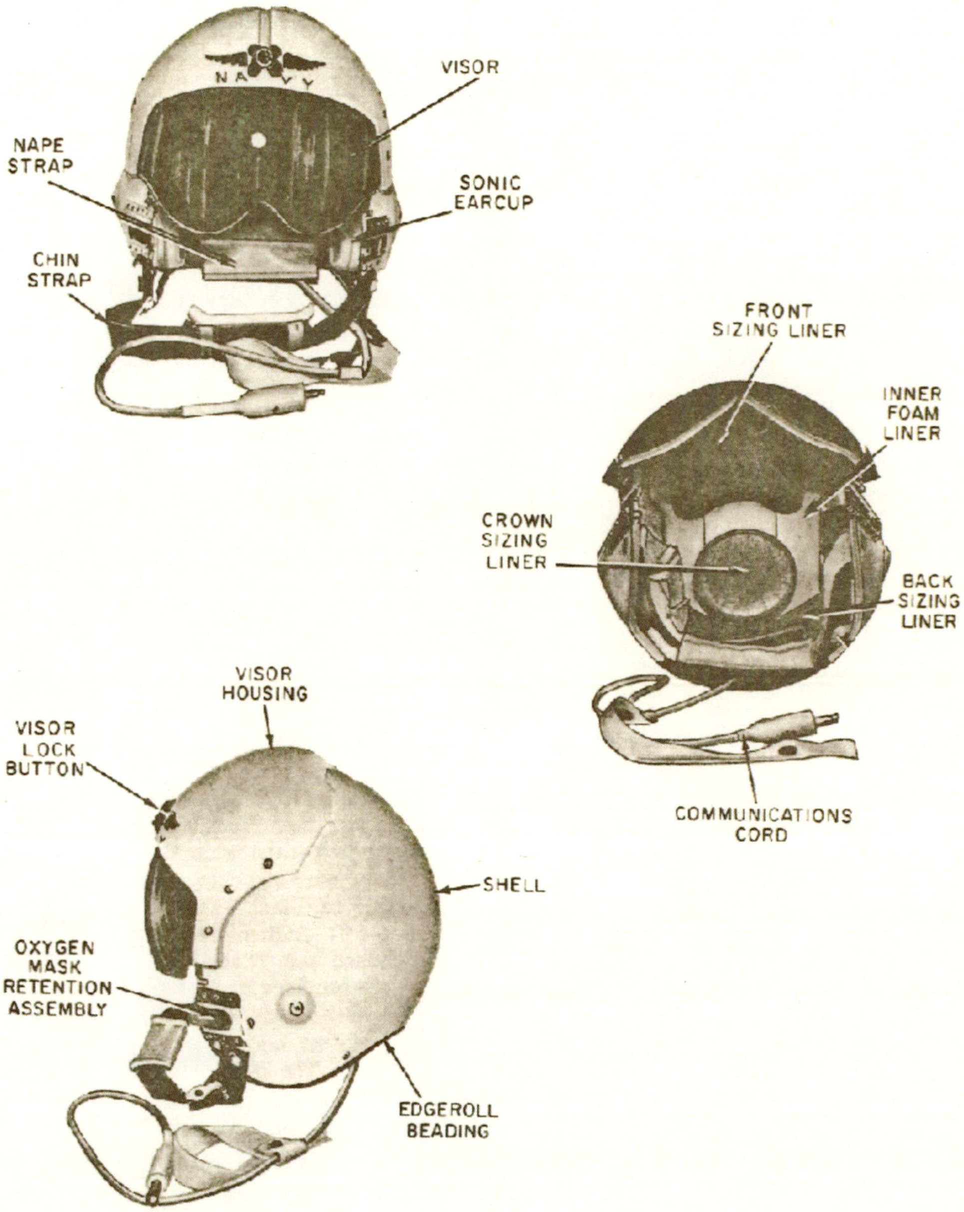

PR.315

Figure 13-1.—APH-6D protective helmet.

(U. S. Navy photo)

The photo above is an AP Wirephoto (author's private collection) dated July 1967. The lucky passenger — Bonnie LeMaster of Marion, Indiana — is wearing an APH-6 flight helmet and, of course, a huge smile as she prepares for her Blue Angels VIP ride. This particular helmet is not equipped with the custom leather edge roll, as was the practice with the pilots' helmets.

This photo is courtesy of PH3 Jim Preston, who served as the Blue Angels staff photographer from 1974 to 1977. Jim is wearing a blue APH-6 flight helmet and is likely to be one of the earliest "selfies."

9
DH-151 Flight Helmet

GENTEX Corporation, through its subsidiary, Protection, Inc., was making custom fit flight helmets from their proprietary shell design, the DH-151, since the early 1970s.

DH-151 is a GENTEX internal definition, "DH" referring to "Designated" or "Design Helmet." The DH-151 story is a bit elusive. We know it exists and has been used extensively in Switzerland, confirmed by GENTEX. We also know that the DH-151 saw U.S. distribution based on several examples of flight helmet identification tags affixed to the inside shell. By many accounts the DH-151 was subsequently standardized as the PRK-37/P and later became the base for most of the U. S. Navy HGU-type helmets and later trimmed at the nape and temples to produce the HGU-55/P.

As discussed in the last chapter, the APH-6 flight helmet and all of its variants, served the U. S. Navy well through the war in Vietnam but, pilot concerns over weight and poor fit prompted a Naval Air Development Center study to address much needed improvements.

The U. S. Navy decided against modifying the APH-6, and instead, evaluated new lightweight helmets sourced from the two major flight helmet suppliers, GENTEX Corporation and Sierra Engineering. GENTEX developed a helmet shell designated DH-151, a few examples which show up in private collections from time to time.

By early 1977 a design was finalized and U. S. Navy contracts were awarded to GENTEX and Sierra for lightweight fiberglass shells, later designated PRK-37/P. These shells were built up into complete helmets with various components based on the requirements of the end user.

According to Steve Norris the "DH" is the nomenclature assigned by GENTEX to their helmets on their work orders. This "DH" designation is the brain child of a former

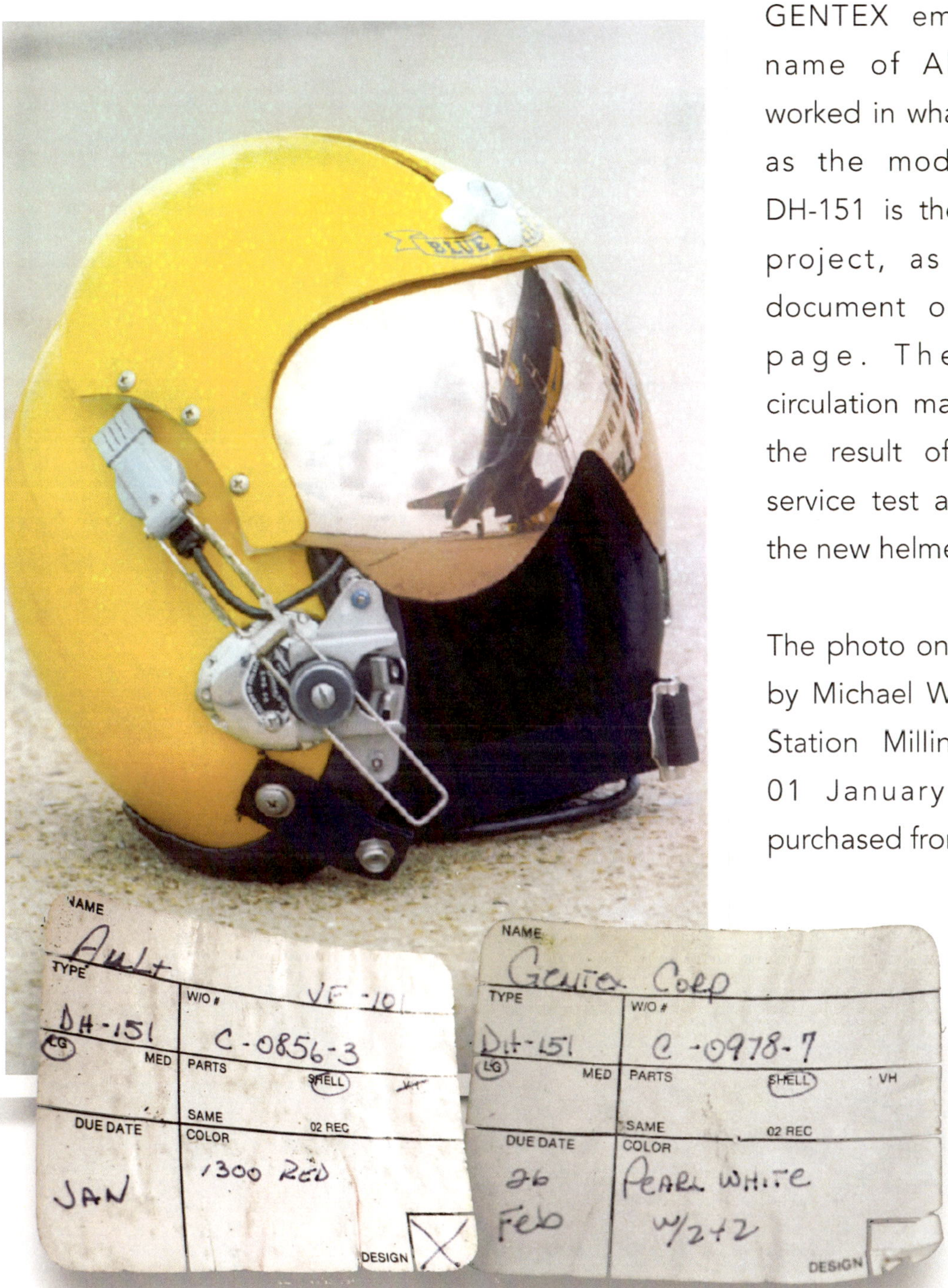

GENTEX employee by the name of Al Kozach, who worked in what was referred to as the model shop. Thus, DH-151 is the 151st assigned project, as shown in the document on the following page. The helmets in circulation marked DH-151 are the result of a GENTEX in-service test and evaluation on the new helmets.

The photo on the left was shot by Michael Wood at Naval Air Station Millington, lennessee 01 January 1979. (Image purchased from Alamy).

The letter on the following page (also courtesy of Steve Norris), dated 12 August 1977, from GENTEX to the Naval Air Development Center, discusses the operational test and evaluation of lightweight aircrew helmets, specifically, the DH-151. The tests were deemed successful, and the DH-151 entered official service with the U. S. Navy as the PRK-37/P (mis-identified in the letter as PRU-37/P). The PRK-37/P shells were the

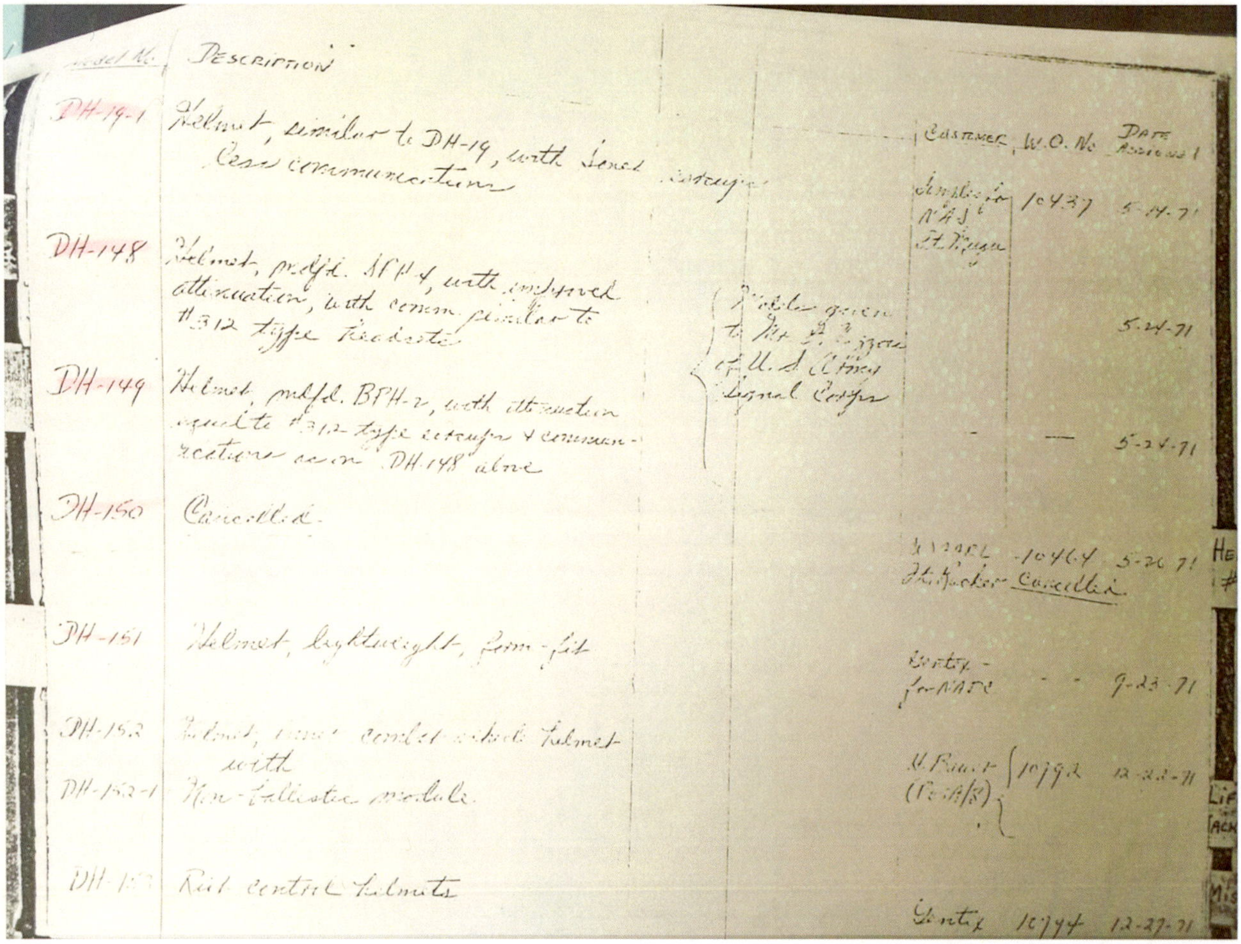

Model No.	Description	Customer	W.O. No.	Date Assigned
DH-19-1	Helmet, similar to DH-19, with [illegible] communications	[illegible]	10437	5-14-71
DH-148	Helmet, modfd. [illegible], with improved attenuation, with comm. similar to #312 type headsets			5-24-71
DH-149	Helmet, modfd. BPH-2, with attenuation equal to #312-type [illegible] & communications as on DH-148 above	-	—	5-24-71
DH-150	Cancelled	[illegible]	10464	5-20-71 Cancelled
DH-151	Helmet, lightweight, form-fit	Gentex - for NATC	-	9-23-71
DH-152	Helmet, [illegible] helmet with			
DH-152-1	Non-ballistic module	[illegible]	10792	12-22-71
DH-15[illegible]	Riot control helmets	Gentex	10794	12-29-71

building block from which a host of U. S. Navy flight helmets were built, including the HGU-33/P and HGU-34/P.

During the test and evaluation period the helmets built from the DH-151 were not equipped with nape straps as the custom, form-fit liners were sufficient to secure the helmet to the wearer's head. It is believed that some helmets may have been subsequently field modified to include a nape strap. When the DH-151 nomenclature changed to PRK-37/P, nape straps became standardized.

It's difficult to determine how many GENTEX flight helmets were built and distributed under the operational test and evaluation period prior to being re-designated the PRK-37/P. Without a physical inspection and verification of an identification tag attached to the inside shell, it is very difficult — if not impossible — to differentiate

DEPARTMENT OF THE NAVY
NAVAL AIR DEVELOPMENT CENTER
WARMINSTER, PA. 18974

404311
6577
12 AUG 1977

Gentex Corporation
Carbondale, PA 18407

Attention: Mr. J. Aileo

Gentlemen:

In response to your inquiry during the telephone conversation between Mr. J. Aileo of your company and Mr. J. Castine of this Center on 21 July 1977, the following documentation applies:

1. Dayton T. Brown, Inc. Test Report No. DTB06R72-1791 entitled, "Comparison of Lightweight Helmet Shells With Form-Fit Liners and With Standard Liner/Sizing Pads," dated 12 December 1972.

2. Commander, Operational Test and Evaluation Force Report on Project 420-OT-III (OPNAV Report Symbol No. 3960-12) entitled, "Operational Test and Evaluation of Lightweight Aircrew Helmets," dated 9 February 1977.

The model DH-151 helmet shell has been designated the PRU-37/P. The U. S. Air Force form-fit liner fabricated in the field from kits has been designated the PRU-39/P and the standard liner/sizing pads has been designated the PRU-40/P. The PRU-37/P helmet shell fitted with the PRU-39/P liner is designated the HGU-33/P helmet assembly and PRU-37/P fitted with the PRU-40/P is the HGU-34/P helmet assembly.

The first report documents the laboratory testing of the HGU-33/P and HGU-34/P to determine the structural integrity and protection characteristics of these lightweight helmet assemblies. Testing consisted of the following:

1. RPHI (Relative Physiological Hazard Index) determination,
2. Windblast integrity (452 knots),
3. Impact - UPC (Ultimate Protective Capability) determination,
4. Penetration - UPC determination,
5. Chinstrap strength, and
6. Slippage test (stability under simulated high G forces).

between a DH-151 and a PRK-37/P based solely on photographs. The lack of nape strap attach holes is certainly a clue, but there is no way to determine if any of these

404311

Results of the above tests were acceptable and operational test and evaluation was recommended. The second report documents the outcome of the operational test and evaluation. The HGU-33/P and HGU-34/P helmet assemblies were issued to operational aircrews and subjective data was recorded on questionnaires. The HGU-33/P was preferred to the HGU-34/P but both were found to be operationally effective, mission compatible, and user acceptable. Approval for Service Use was recommended. It is anticipated that the HGU-33/P and HGU-34/P helmet assemblies will become Navy support items during 1979. The Navy has used the DH-151 helmet shell as a platform for the R-1826 AVG-8 VTAS (Visual Target Acquisition System) since 1972. This configuration with the PRU-39/P liner is designated the HGU-37/P.

Your continued interest in military life support equipment is encouraged.

J. HARDING
By direction

DH-151 helmets were field modified to include a nape strap, making them indistinguishable from their PRK-37/P cousins.

Since the test and evaluation period for the DH-151 began in 1971 and was re-designated PRK-37/P in 1977 we can surmise that the life of the DH-151 falls within this 6 year period.[1]

The screen shot on the following page is courtesy of Ryan Nothhaft, grandson of CAPT Butch Voris, who is dedicated to celebrating the rich history of the U. S. Navy Blue Angels through the production of numerous video interviews of former Team members.[2]

This DH-151 helmet belongs to Lt. Col. (then Captain) Kevin O'Mara, USMC[3], who flew the #2 Right Wing for the Blue Angels in 1970 and 1971. I reached out to Ryan

[1] This reference is for the U.S. only; The DH-151 designation was also used for some overseas sales.

[2] https://www.youtube.com/c/blueangelphantoms

[3] Lt. Col. Kevin O'Mara, USMC (Retired)

Nothhaft and he shared with me that during his interview, Lt. Col. Kevin O'Mara, informed him that he was responsible for the Blue Angels adopting the gold anti-nuclear flash visors. (Inset photo is courtesy of Lt. Col. O'Mara).

In Ryan's interview, Kevin O'Mara said he reported to NAS Pensacola in November of 1969 to begin his training with the U. S. Navy Blue Angels. The show season was coming to an end and O'Mara was able to fly in the back seat of Blue Angels #2, piloted by Capt. Vince Donile, USMC enroute to an air show in El Paso, Texas. The Team experienced extreme head winds — much greater than had been forecast. The problem grew increasingly dire, and they were quickly face-to-face with their worst fears: fuel starvation.

Just short of the field in El Paso, Donile and O'Mara were left with no other options, and forced to eject. Lt. Col. O'Mara suffered a triple compression fracture, which typically results in a 3-month suspension from flying. After a convincing plea with the flight surgeon at NAS Pensacola, O'Mara was issued a waiver so he could continue training with the Blue Angels.

U. S. Navy photo.
(Author's private collection)

As fascinating as this story was, I was curious about Lt. Col. O'Mara's claim of introducing the Blue Angels to the gold, nuclear anti-flash visor. Through a series of contacts, I located, and was able to speak with Col. O'Mara.

According to Col. O'Mara, he flew his first Vietnam combat tour in the F-4 "Phantom" between 1967 and 1968 out of Marine Corps Air Station Chu Lai, about 50 miles south of Danang. Col. O'Mara said, "We flew sorties 24 hours a day, and our helmets were fitted with dual visors, dark for daytime and clear for night sorties." He complained that the dark visor was just not up to the job, so reached out to a friend flying A-6 "Intruders" — also out of Chu Lai — for assistance. "The A-6 crews were outfitted with a unique gold visor because they had some special missions[4], and he agreed to get me one. Needless to say, I found the new gold visor exactly what I had been looking for."

Col. O'Mara continues, "When I was selected for the Blue Angels in 1970, we began flying out of NAF El Centro, CA for 2 1/2 months of winter training. I flew the #2 position — the right wing in the Diamond formation. The Diamond launched twice each morning, seven days a week. Often on the first hop the sun angle was so low that flying close formation was a real challenge visually. I suggested we try the A-6 gold visors and everyone agreed it was a great improvement. That was the beginning of the Blue Angels gold visors."

"That was the beginning of the Blue Angels gold visors."

Lt. Col. Kevin O'Mara
NAF El Centro, CA
1970

Little did Col. O'Mara know at the time, what began as a flight safety issue immediately became an aesthetically pleasing addition to their flight helmets that has stood the test of time for over 50 years! The gold visors are so iconic that the "look" caught the eye of their sister organization in the U. S. Air Force. In 2020, the USAF Thunderbirds tested silver mirrored visors. The Team loved the new appearance and unanimously voted to switch to the silver visors beginning with the 2021 show season. They say imitation is the truest form of flattery!

These early DH-151 flight helmets feature nickel plated oxygen mask receivers from the Hardman Tool & Engineering Company.

[4] The gold coated visors were issued to flight crews flying aircraft with nuclear capability. The gold coating darkened the visor and was designed to reflect the potentially blinding effects of a nuclear blast, providing the crew members' eyes some degree of protection.

The photo above is from the 1971 show season. (U. S. Navy Blue Angels photo). The 1971 Blue Angels pose in front of their F-4J "Phantom. Left to right are, Dick Schram (#8 - Advance Pilot), Bill Switzer (#6 - Opposing Solo), Jim Maslowski (#4 - Slot), Kevin O'Mara (#2 - Right Wing), Harley Hall (#1 - Boss / Commander), Bill Beardsley (#3 - Left Wing), Skip Umstead (#5 - Lead Solo) and J. D. Davis (#7 - Narrator).

On the next page is a GENTEX worker's mark on the inside shell of a DH-151, size Large. Collectors know all too well the importance and value of peeking behind the helmet liner and ear cups for manufacturer's tags and maker's marks. Unfortunately, not all helmets reveal their identity so easily. Many are void of any identifying marks, making a positive identification challenging and all to often frustrating.

On the following pages, we see another historical document referring to the GENTEX DH-151, as well as Sierra Engineering's prototype, 570-1301. This document — Form DD-61 - Request For Nomenclature[5] — details the U. S. Navy's intent to formally assign an alpha-numeric designation for these prototype helmets.

It is interesting and curious to note that this 1973 dated DD-61 originated from the Naval Air Systems Command, routed through the Naval Air Engineering Center to the Commander, Headquarters, Aeronautical Services Division (AFSC), Wright Patterson Air Force Base. Why was the U. S. Navy submitting a request for U. S. Air Force approval? Here's a little background — compliments of Steve Norris — to help the uninformed reader (of which I was one!) understand this complicated military procurement process.

[5] From the personal collection of Steve Norris.

1. ORIGINATOR & ADDRESS
Commander
Naval Air Systems Comma...
Washington, D. C. 20361

REQUEST FOR NOMENCLATURE

2. THRU OR VIA
ESSD
Naval Air Engineering Center
Philadelphia, Pennsylvania 19112

3. TO:
Commander, ..., ASD (AFSC)
Wright-Patterson AFB
Ohio 45433

4. DATE OF REQUEST	5. DESCRIPTION PER DP NO.	6. SOURCE REQUEST NO.	7. SECURITY CLASS OF EQUI
5-11-73	Figure 1	NA-73-4065	Unclassified

8. FEDERAL SUPPLY CLASS: 8475
9. STOCK NO. (When available)
10. ACTION REQUESTED: ☐ REVISION ☐ CANCELLATION ☒ ASSIGNMENT

11. FOR REVISIONS NOTE CHANGE IN: ☐ ITEM NAME ☐ TECHNICAL DATA ☐ TYPE DESIGNATION ☐ SECURITY CLASS OF EQUIP ☐ SECURITY CLASS OF TECH DATA

12. TYPE OF NOMENCLATURE REQUESTED (Check one): ☐ EXPERIMENTAL OR DEVELOPMENTAL ☒ PREPRODUCTION OR PRODUCTION

13. RECOMMENDED NOMENCLATURE
Helmet, Pilot's Tactical HGU-33/P E( )

TECHNICAL DATA

14. (1) FEDERAL CATALOGING ITEM NAME

1. Helmet, Pilot's Tactical
2. Lightweight Fighter/Attack Pilot's High "g" Helmet
4. Weight: 2.25 lb./ 2.35 lb.
   Dimensions: Fits in a 19 inch sphere
6. a. Epoxy impregnated fiberglass cloth helmet shell
   b. Polyurethane leather covered foam liner
   c. Epoxy impregnated fiberglass cloth or plastic visor housing
7. Is a composite of the following:
   a. Gentex Corporation part # DH-151
   b. Sierra Engineering part # 570-1301
   c. USAF On-Site Molding System Liner
8. On-Site Custom form fitted liner to wearer's head

11. (a and b) One of the following:

Gentex Corporation
Carbondale, Pennsylvania 18407
P/N (TBA)
Sierra Engineering Company
123 East Montecito Avenue
Sierra Madre, California 91024
P/N (TBA)

d) MIL-P-83379
AS 3721
AS 3731

15. FUNCTIONAL DESCRIPTION
Lightweight tactical pilot's high "g" protective helmet in two sizes—— MEDIUM and LARGE (form fitted)

16. CONTRACT OR ORDER NO.	17. GOVT DRAWING NO.	18. GOVT SPECIFICATION NO.
		AS 3721/MIL-P-83379/AS 3731

19. DATE ACTION TAKEN TO (For use by Control Point only): ☒ ASSIGN 12 June 1973 ☐ CANCEL ☐ REVISE
20. PROJECT GROUP: AIR-5311J

21. EQUIPMENT OF WHICH THIS ITEM IS A PART
Personal Issue Helmet

22. EQUIPMENT WITH WHICH THIS ITEM IS USED
Tactical Fixed Wing Aircraft with Automated Escape Systems

DD FORM 1 APR 61 61 REPLACES PREVIOUS EDITIONS OF THIS FORM, WHICH ARE OBSOLETE.
ENCLOSURE

NA-73 406.5

23. INITIATOR REQUESTING SUFFIX LETTER [illegible] ASSIGNMENT OR NEW ASSIGNMENT WILL CHECK [illegible] APPROPRIATE BLOCK. COMPLETE DETAILS CONCERNING SIMILARITIES, [illegible] FERENCES, AND INTERCHANGEABILITY WILL [illegible] STATED IN "OTHER PERTINENT INFORMATION" BLOCK BELOW.

- ☐ TWO WAY INTERCHANGEABLE, EXCEPT BY MAINTENANCE PARTS, WITH *(List equipments)*
- ☐ TWO WAY INTERCHANGEABLE, INCLUDING MAINTENANCE PARTS, WITH *(List equipments)*
- ☐ ONE WAY INTERCHANGEABLE WITH *(List equipments)*
- ☐ SIMILAR TO ______________ *(List equipments)* BUT NOT ☐ ELECTRICALLY, ☐ MECHANICALLY, ☐ FUNCTIONALLY, INTERCHANGEABLE *(Check appropriate block or blocks and specify differences)*

N/A

24. OTHER PERTINENT INFORMATION *(List any additional information not covered by the above questions concerning function, application, purpose, relationship or similarity to other equipment, reason for revision, substitutability of or by other equipments, description of the design change, etc., which would aid in the assignment of nomenclature to this request.)*

N/A

R. SANTEE, ENG. TECH.
TEL. 215-755-3892

Ray Santee

25. INITIATED BY *(Name, Title & Telephone Extension)*	26. SIGNATURE
Glenn M. Smith, Project Engineer 692-7480	Glenn M. Smith

FOR USE BY NOMENCLATURE CONTROL POINT ONLY

27. AUTHORIZED NOMENCLATURE SEE BLOCK 13&19

28. AUTHORIZED BY *(Name, Title & Telephone Extension)*	29. SIGNATURE
F. LOTSPAITH Type Designation Branch Directorate of Engineering Standards	Mary L. Robinson

A15578

REQUEST FOR NOMENCLATURE

1. ORIGINATOR & ADDRESS
Commander
Naval Air Systems Command
Washington, D. C. 20361

2. THRU OR VIA
ESSD
Naval Air Engineering Center
Philadelphia, Pennsylvania 19112

3. TO
Commander, HQS., ASD (AFSC)
Wright-Patterson AFB
Ohio 45433

4. DATE OF REQUEST	5. DESCRIPTION PER DP NO.	6. SOURCE REQUEST NO.	7. SECURITY CLASS OF EQUIP
5-11-73	Figure 1	NA 73-4066	Unclassified

8. FEDERAL SUPPLY CLASS: 8475
9. STOCK NO. (When available):
10. ACTION REQUESTED: ☐ REVISION ☐ CANCELLATION ☒ ASSIGNMENT
11. FOR REVISIONS NOTE CHANGE IN: ☐ ITEM NAME ☐ TECHNICAL DATA ☐ TYPE DESIGNATION ☐ SECURITY CLASS OF EQUIP ☐ SECURITY CLASS OF TECH DATA
12. TYPE OF NOMENCLATURE REQUESTED (Check one): ☐ EXPERIMENTAL OR DEVELOPMENTAL ☒ PREPRODUCTION OR PRODUCTION
13. RECOMMENDED NOMENCLATURE
Helmet, Aircrew Protective — HGU-34/P ( )

TECHNICAL DATA

14. (1) FEDERAL CATALOGING ITEM NAME

1. Helmet, Aircrew Protective
2. Lightweight Pilot's Protective Helmet
4. Weight: 2.25 lb.
   Dimensions: Fits in a 19 inch diameter sphere
6. a. Epoxy impregnated fiberglass cloth helmet shell
   b. Liner--Polyurethane foam (leather covered)
   c. Visor Housing--epoxy impregnated fiberglass cloth or plastic
7. Is a composite of the following:
   a. Gentex Corporation Part # DH-151.
   b. Sierra Engineering Part # 570-1301
   c. USAF/Navy fitting system
8. Lightweight pilot's helmet with pressure sensitive top, front, back and temporal fitting pads

11. One of the following: (a and b)

Gentex Corporation
Carbondale, Pennsylvania 18407
P/N (TBA)
Sierra Engineering Company
123 East Montecito Avenue
Sierra Madre, California 91024
P/N (TBA)

d) AS 3721
AS 3723

15. FUNCTIONAL DESCRIPTION
Lightweight nontactical pilot's helmet with universal pressure sensitive fitting pads, in two sizes— MEDIUM and LARGE

16. CONTRACT OR ORDER NO.	17. GOVT DRAWING NO.	18. GOVT SPECIFICATION NO.
		AS 3721/AS 3723

19. DATE ACTION TAKEN TO (For use by Control Point only): ☒ ASSIGN 12 June 1973 ☐ CANCEL ☐ REVISE
20. PROJECT GROUP: AIR-5311J
21. EQUIPMENT OF WHICH THIS ITEM IS A PART
Personal Issue Helmet
22. EQUIPMENT WITH WHICH THIS ITEM IS USED
Nontactical fixed wing aircraft

DD FORM APR 61 61 — REPLACES PREVIOUS EDITIONS OF THIS FORM, WHICH ARE OBSOLETE. ENCLOSURE

The simple explanation is that in 1952 the Defense Supply Agency (DSA) — forerunner to the Defense Logistics Agency (DLA) — was created to combat procurement redundancy within the various branches of military service organizations. For the purpose of this story all requests for new flight helmets were routed through the Aircrew Systems Directorate (ASD) at Wright Patterson AFB — the designated authority — for approval. This included requests from the U. S. Air Force, U. S. Navy and U. S. Army. This was the theory. In reality, most U. S. Government-run bureaucracies tend to go a little sideways, and this was no exception. But that's a long story, for another time.

I digress. Now, back to the story.

The U. S. Navy was proposing that these lightweight, tactical pilots, high "g" protective helmets, intended for tactical, fixed wing aircraft, with Automated Escape Systems be assigned the nomenclature, **Helmet, Pilot's Tactical, HGU-33/P.** [6]

The request also specified a lightweight, non-tactical, pilot's helmet with pressure sensitive fitting pads, intended for non-tactical fixed wing aircraft and be assigned the nomenclature, **Helmet, Aircrew Protective, HGU-34/P.**

We know from documents posted at the beginning of this chapter that the DH-151 went into full production — re-designated the PRK-37/P shell — that was the basis for building the U. S. Navy HGU-33/P, HGU-34/P and many more variants.

Serious collectors of military flight helmets — and I am a card-carrying member — are all too often faced with identifying the age or era of a particular item. To make matters worse, reference photos that can clearly be dated, sometimes feature flight helmets that do not seem period correct.

The likely explanation is the early HGU-33/P-type flight helmets (built up from the PRK-37/P shell) were initially made in the style of the DH-151.

[6] HGU stood for Helmet, General Utility.

Blue Angels DH-151 (early example) replica in author's personal collection. (Author's photo)

Anomalies in the flight gear collecting arena have been, and continue to be challenging to sort out. If we are lucky, some offer reasonable explanations, such as personal preferences and supply availability. Others simply defy explanation.

LT Bill "Burner" Beardsley, (below) Left Wing #3, circa 1971 wearing his DH-151 flight helmet. The silver metal flake is the "signature" of custom flight helmets made for the Team by Protection, Inc., a former subsidiary of GENTEX. Life Support folks and savvy collectors will immediately notice the missing screw on the visor lock knob.

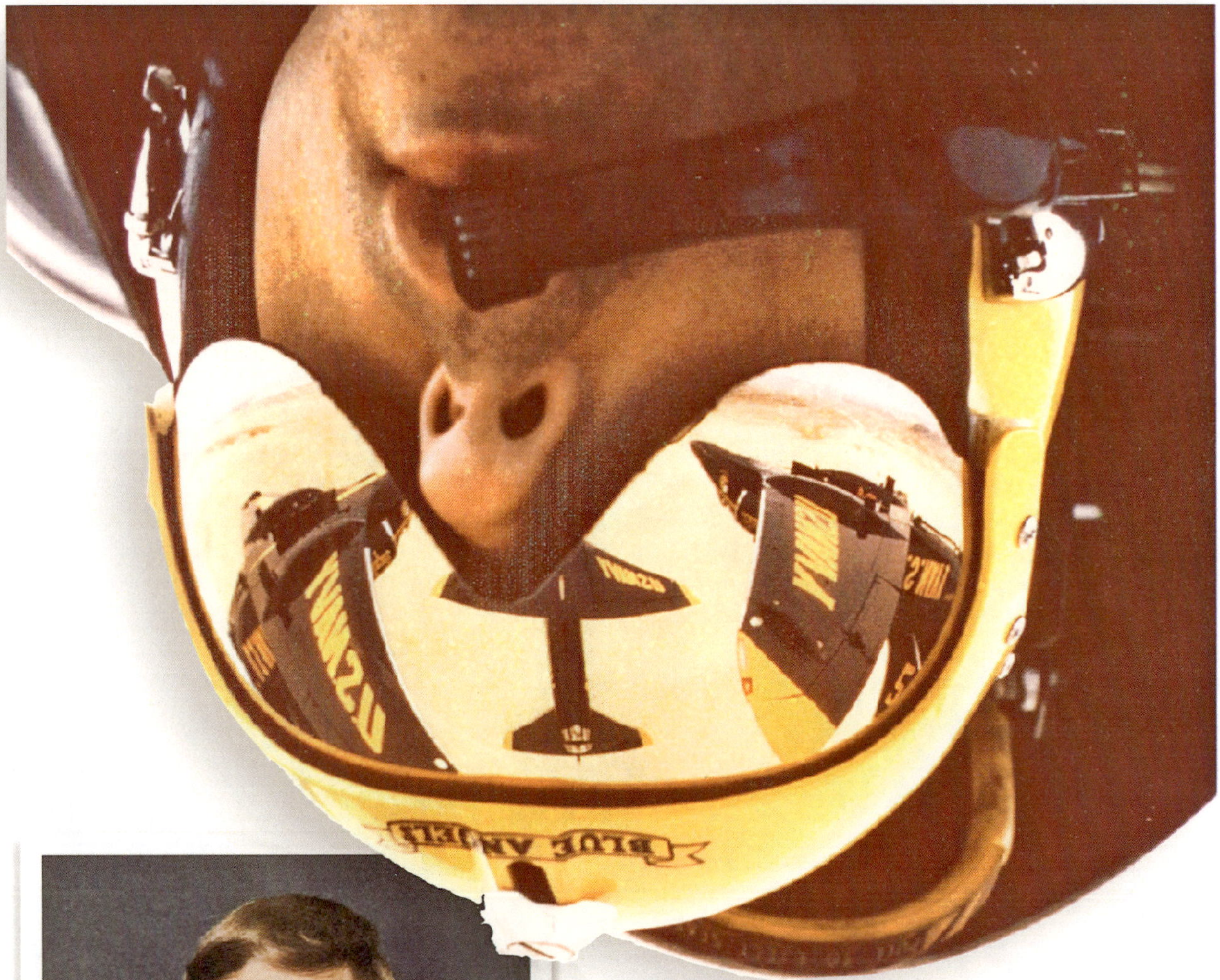

The photo above is of Blue Angels slot pilot John Chehansky from 1975. The flight helmet is a DH-151 with Hardman Tool & Engineering Company oxygen mask receivers. This photo was captured by PH3 Jim Preston, Blue Angels Staff Photographer, 1974 - 1977. Jim believes this was one of the very first times a camera was remotely mounted in the cockpit of a numbered demonstration jet.

On the left is a photo of a young Jim Preston from his service time with the Blue Angels.

The DH-151 Blue Angels flight helmet photos on this page belonged to CDR Harley Hall. and are courtesy of Ryan Nothhaft[7] The helmet is on display at the Harley Hall Building, Seattle, Washington.

CDR Harley Hall was the Blue Angels "BOSS" for the 1970 and 1971 show seasons. In 1973, CDR Hall[8] was the Executive Officer of VF-143 *"Pukin Dogs"* flying off of the *USS Enterprise*. He was shot down over north Vietnam while piloting an F-4J on the last day of official American hostilities in Vietnam. Sadly, CAPT Hall was captured and presumably died in captivity as a prisoner of war; however, a forensic examination of CAPT Hall's remains by his family tended to raise more questions than answers.

Rest In Peace, Sir.

---

[7] blueangelphantoms.com.

[8] Posthumously promoted to Captain.

The quality of this photo is poor at best, but still serves as early evidence of when the Blue Angels crest began to appear on the rear of the helmet. In this photo, CDR Marvin Francis "Skip" Umstead — who served as "Boss" of the Team for the 1973 show season — is the earliest example thus far showing the use and placement of the crest. Earlier in this chapter, Capt. Kevin O'Mara's helmet from the 1970 and 1971 show seasons was void of the crest. It also appears that the entire crest, with the Blue Angels "scroll" was used. Tragically, CDR Umstead was killed when his jet collided with another during a rehearsal over Lakehurst Naval Air Station, New Jersey in July of 1973. (Photo submitted by John Slezak)

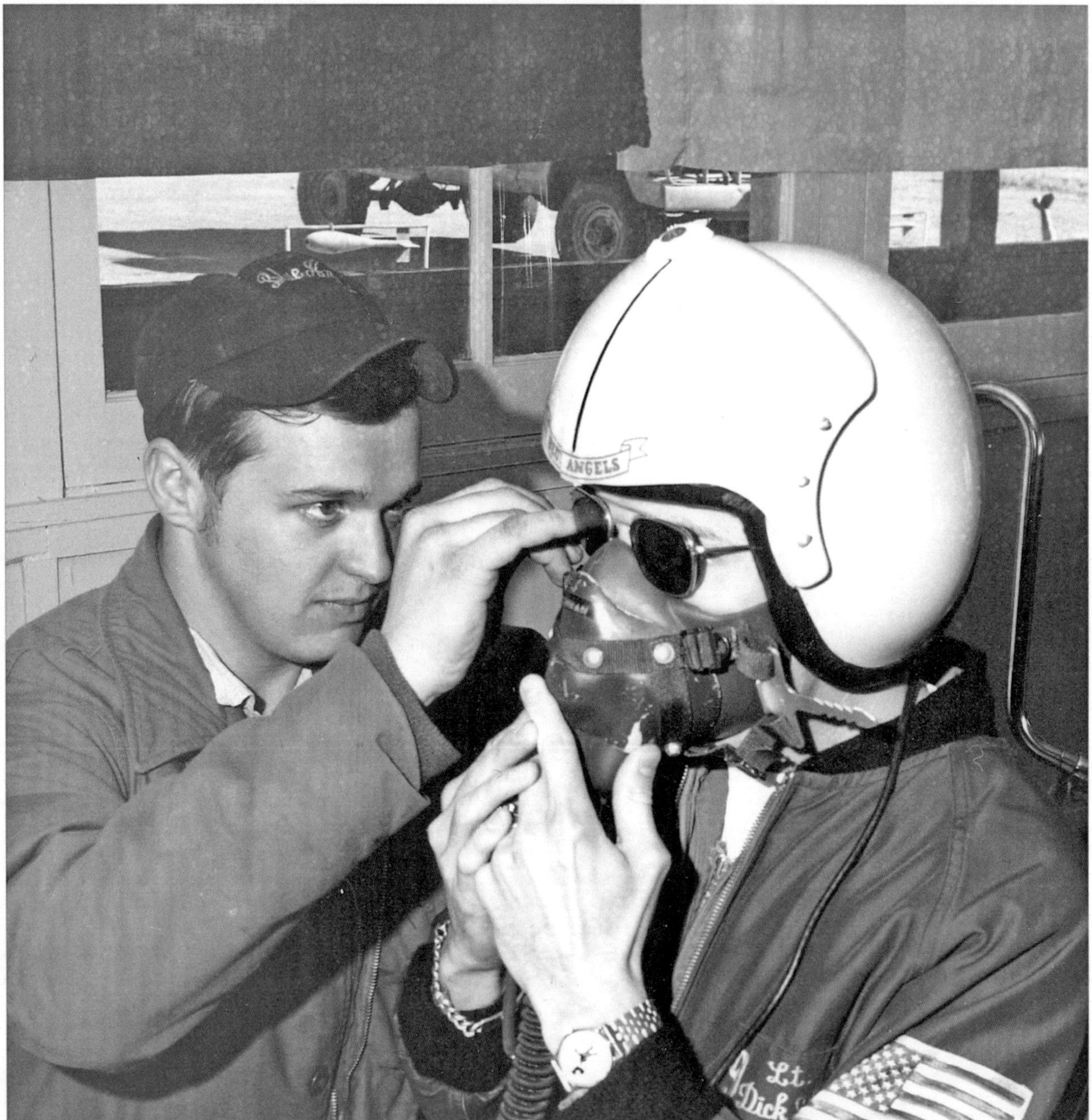

The previous chapter featured a beautiful example of LT Dick Schram's APH-6 flight helmet from 1969. Above, LT Schram is getting fitted by Dave Shultz for a new DH-151 flight helmet. The MS-22001 oxygen mask is placed in the correct position on the pilot's face, and the helmet shell marked for the location of the Hardman Tool & Engineering Company oxygen mask receivers. (U. S. Navy Blue Angels photo courtesy of Richard Keane, Blue Angels Alumni Association)

The photos on this and the following page (author's photos) are of the same helmet issued to LT Dick Schram as seen on the previous page.

This beautiful and rare example is from the private collection of John Slezak. The clear coat is impregnated with silver metal flake, a tell-tale sign of Protection, Inc., the company responsible for the customization of the Blue Angels flight helmets during this time period (see Chapter 20).

The back of the helmet features the name "Dick" in white vinyl, just like the example of Kevin O'Mara's helmet from the same era seen earlier in this chapter.

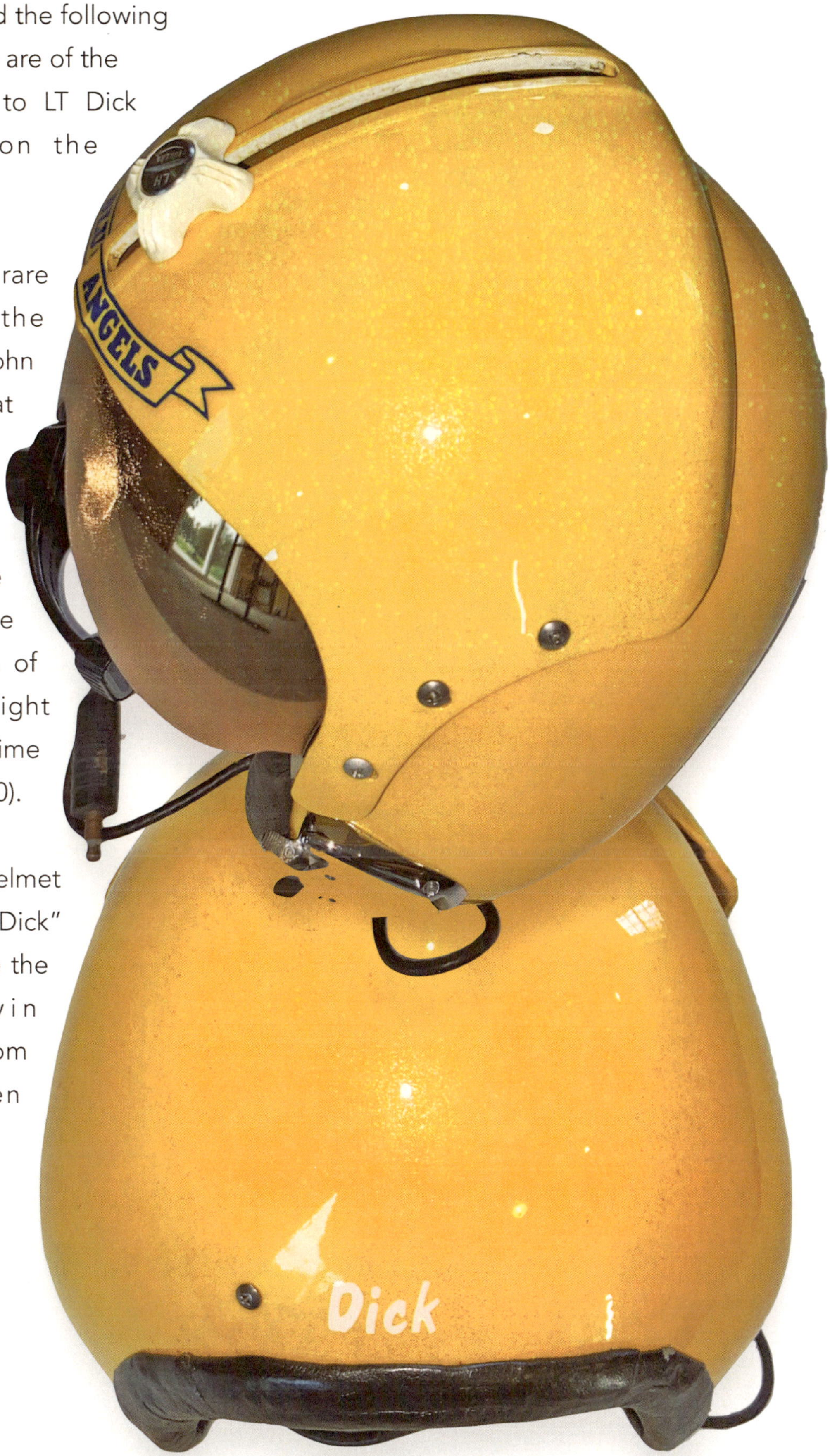

The detailed photos on this page show the nickel plated Hardman Tool & Engineering Company oxygen mask receivers installed by Dave Shultz.

This is one of the very first Blue Angels flight helmets to feature the iconic gold visor. We have Kevin O'Mara to thank for this stunning feature that has withstood the test of time.

Self-portrait of Jim Preston, by Jim Preston, U. S. Navy Blue Angels Staff Photographer, 1974 - 1977. Jim is wearing a blue DH-151 flight helmet with Hardman Tool & Engineering Company oxygen mask receivers. Today, Jim is the Supervising Photographer for the Smithsonian Institution, National Air & Space Museum.

Photo by Jim Preston, U. S. Navy Blue Angels Staff Photographer, 1974 - 1977.

10
HGU-33/P Flight Helmet

By early 1977 the DH-151 design was finalized, approved and U. S. Navy contracts were awarded to GENTEX and Sierra for lightweight fiberglass shells with the new designation PRK-37/P. These shells were built up into complete helmets with various components based on the requirements of the end user. By 1980, U. S. Navy procurement was well underway.

The requirements of the Blue Angels were similar to their counterparts in the fleet — fast jet, high altitude, custom fit liner, etc. The components necessary to fulfill these requirements, paired with the PRK-37/P shell, received the U. S. Navy designation HGU-33/P.

The HGU-33/P configured flight helmets included a single visor housing kit, designated EEK-4A/P, shown on the right.

A GENTEX brochure from the 1980s describes the HGU-33/P as "A refinement of the Navy APH-6 helmet

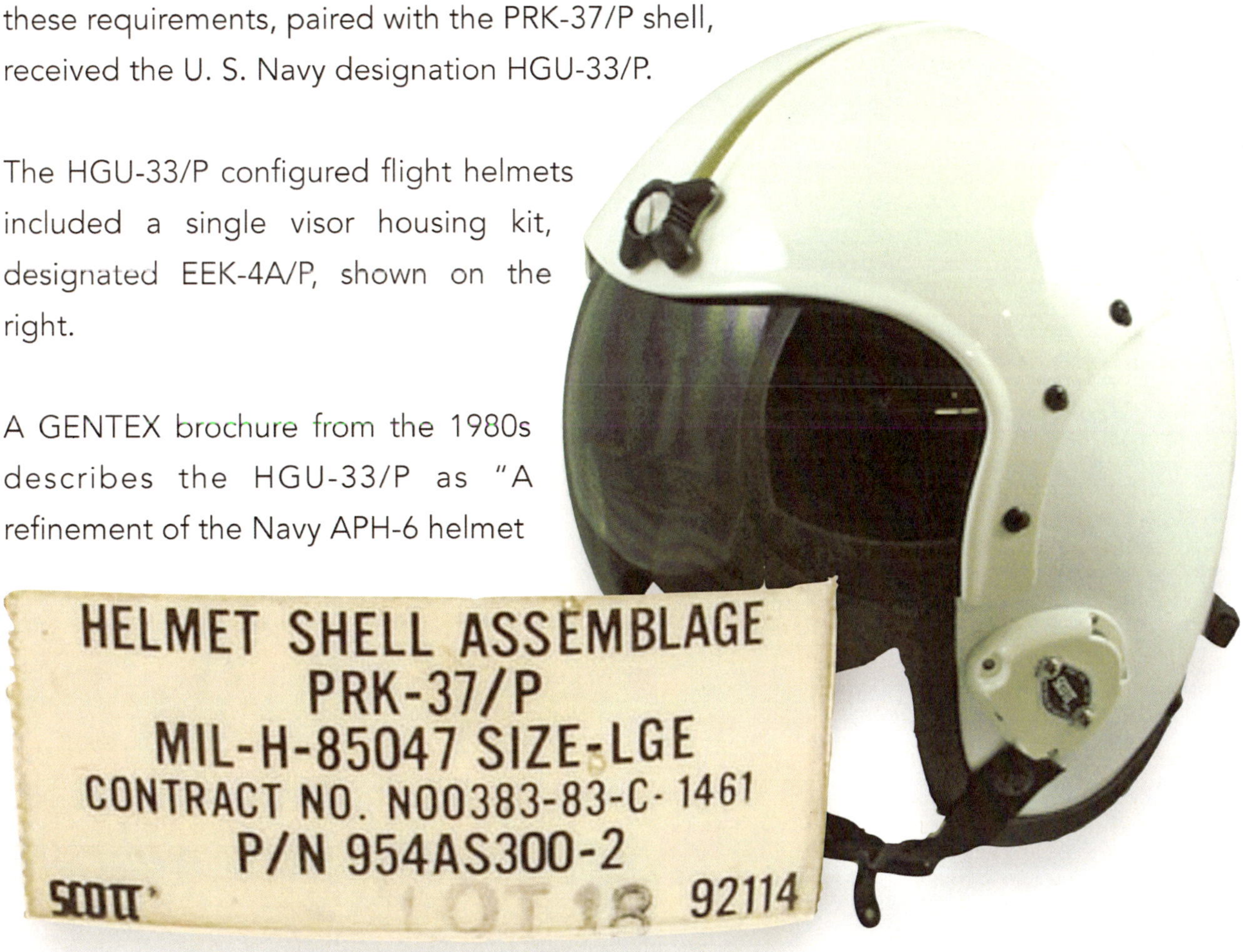

in which advanced lightweight materials enable aircrewmen to comfortably perform as required by current attack and fighter mission envelopes."

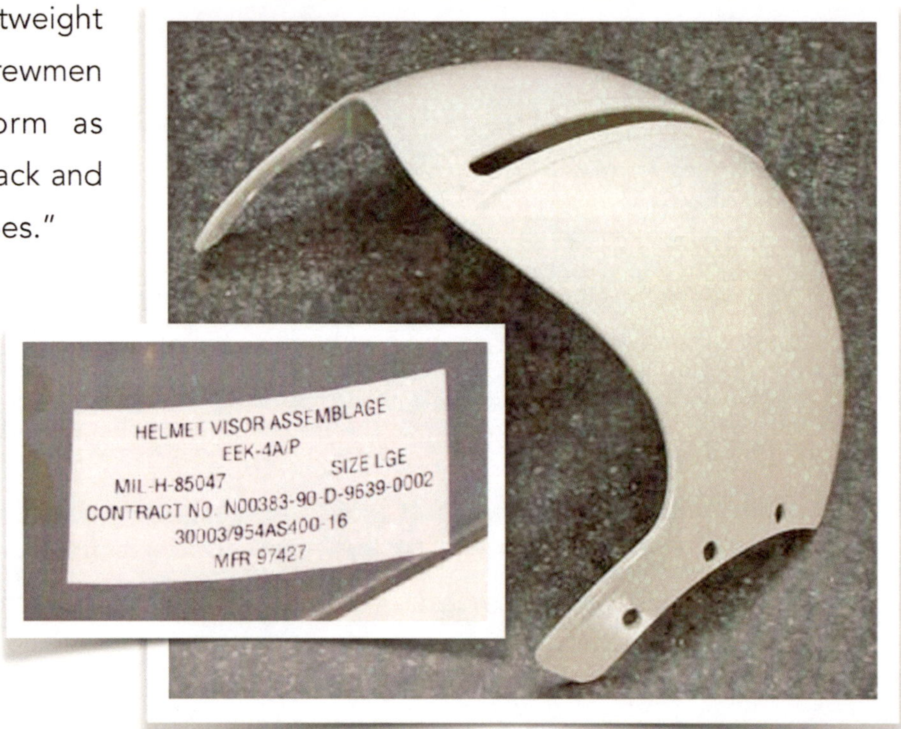

But, there's more to the story. According to James Wegge, Director of Gibson & Barnes[1], the Blue Angels' flight helmets during this period used slightly different components than a standard issue U. S. Navy HGU-33/P therefore, are technically *not* HGU-33/Ps. They are based off of the PRK-37/P shell and used many components common to the HGU-33/P. Without question, an often debated topic among flight gear collectors across the globe. To further complicate this matter Gibson & Barnes adopted and assigned the designation HGU-33/P to all of their PRK-37/P-based helmets sold in the civilian and law enforcement markets. This has led to many debates — some ongoing — in the collecting world.

HGU-33/P

[1] Gibson & Barnes (formerly doing business as Flight Suits, Ltd.) has been the single source provider of custom flight helmets to the Blue Angels since 1984.

"A refinement of the Navy APH-6 helmet in which advanced lightweight materials enable aircrewmen to comfortably perform as required by current attack and fighter mission envelopes."

GENTEX

The helmet on this page belonged to LCDR Mike Gershon[2], who flew with the Blue Angels as the Narrator (#7) and Opposing Solo (#6) for the 1984 and 1985 show seasons. This helmet resides in archives at the National Museum of Naval Aviation at NAS Pensacola (author's photos).

This helmet — like so many others — has several stories to tell, but also harbors many unanswered questions.

The visor housing bears the trademark Blue Angels "scroll" or "banner" below the visor

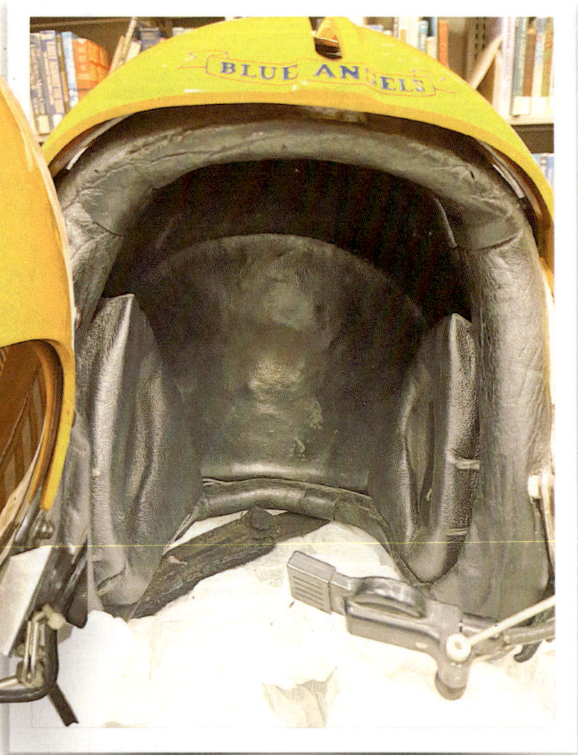

2 Killed during an air show on 13 July 1985.

track that is cut from commercially produced decals like the one shown on the right (from author's collection). The addition of fine, silver metal flake in the finish is a tell tale sign of the work produced by Protection, Incorporated. It is interesting to note that the helmet shell is void of any silver metal flake.

On the rear of the helmet the Blue Angels crest — typically cut from the same decal sheet as the scroll / banner — presents some anomalies that defy explanation. The Naval Training Command "disk" positioned on the lower left hand side of the crest appears as if it could have been cut from the decal sheet. The rest of the crest, on the other hand, has the appearance of being hand painted and showing considerable wear.

"As an aviator, he was the best – a professional's professional, a Blue Angels' Blue Angel."

CDR Larry "Boss" Pearson

LCDR Gershon was tragically killed at an air show at Niagara Falls International Airport when his jet collided with another jet piloted by LT Andy Caputi, who ejected and parachuted to safety.

At LCDR Mike Gershon's funeral, Blue Angels "Boss" CDR Larry Pearson was quoted, "As an aviator, he was the best — a professional's professional, a Blue Angels' Blue Angel."

Rest in Peace, Sir.

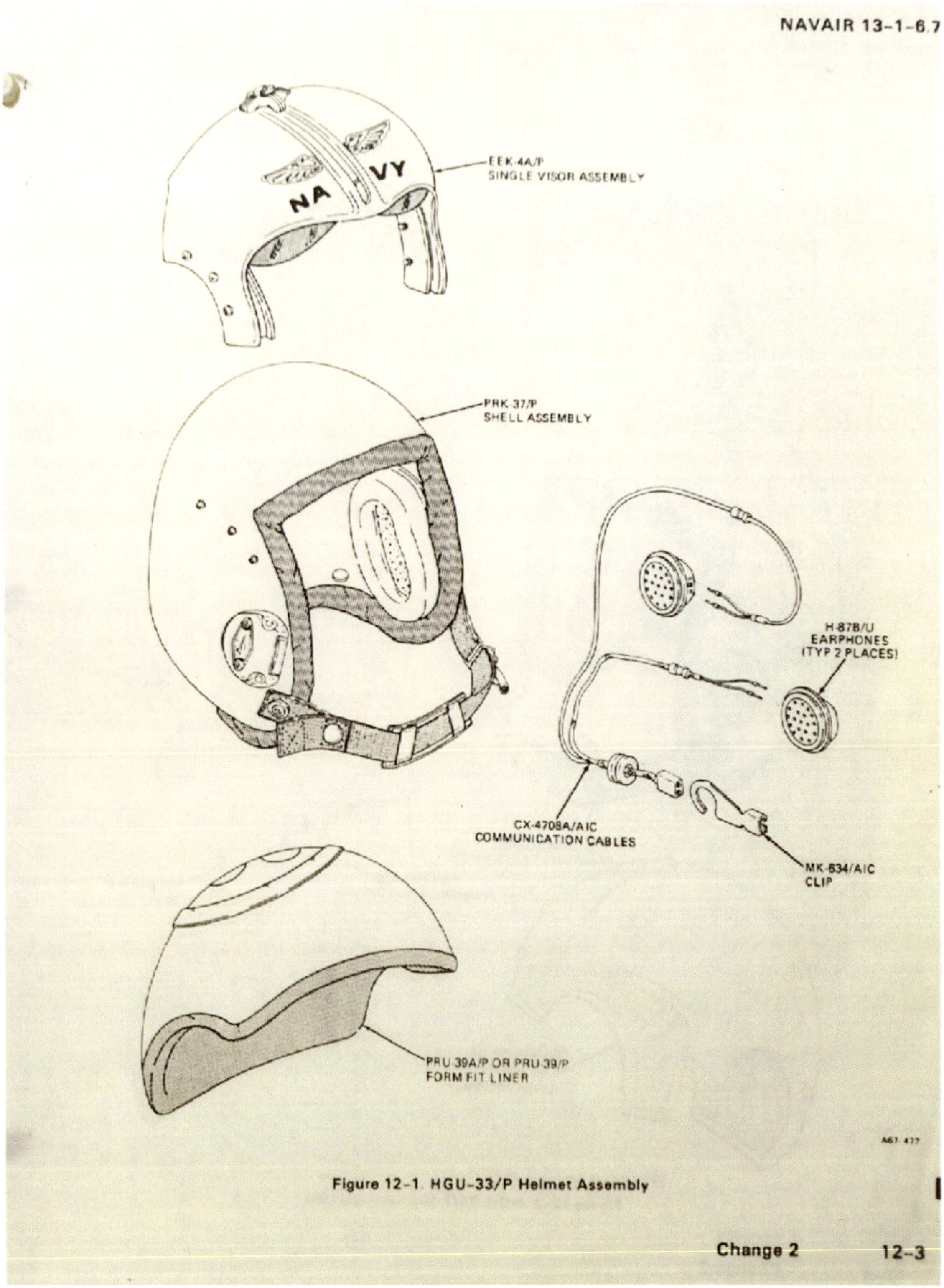

This is a page from NAVAIR 13-1-6.7, Aircrew Personal Protective Equipment, dated 15 August 1979 showing the various components that make up the U. S. Navy HGU-33/P flight helmet (Steve Norris Collection).

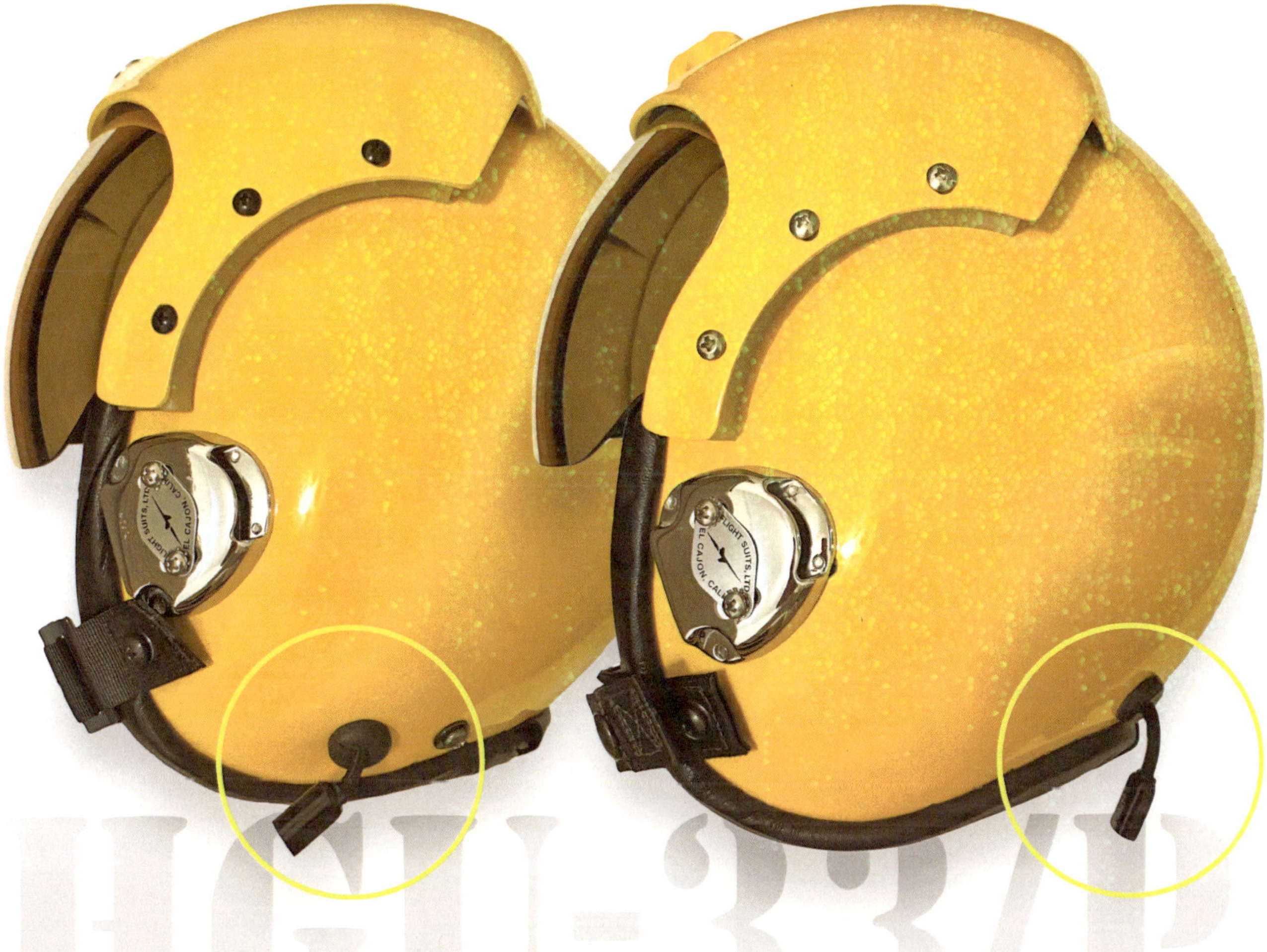

Above is a side-by-side, compare and contrast visualization of the communications cord and grommet locations on an early HGU-33/P[3] (right), which is located far aft and void of a nape strap, and a more contemporary HGU-33/P (left) which is forward of the nape strap screw and includes a nape strap. (Author's private collection).

The HGU-33/P-style helmets enjoyed a long history with the Blue Angels, by some estimates, about 20 years[4]. During this period the helmet remained virtually unchanged, rather unremarkable for a critical life support component. Despite the introduction of the HGU-55/P flight helmet — introduced in 1982 — the Blue Angels continued to use the HGU-33/P-style flight helmets for an additional 10-15 years for reasons that are not clear. A likely explanation will be discussed in Chapter 11 HGU-55/P Flight Helmet.

---

[3] HGU-33/P "style" helmet built from the PRK-37/P shell.

[4] 1977 - 1997.

There is another example of an early HGU-33/P-style Blue Angel Flight Helmet on display at the Aviation Museum of Kentucky in Lexington.

The helmet belonged to LCDR Mike "Nordo" Nord, who flew with the Team in 1980 (photos this page). The visor housing and shell bear the telltale calling card of Protection, Inc. — silver metal flake in the clear coat. This would have been near the end of PI's reign prior to being sold to Flight Suits, Ltd. in 1983. Flight Suits, Ltd., now doing business as Gibson & Barnes, kept one PI employee for continuity with the Blue Angels contract, which was part the buyout agreement. (Author's photos).

Below is another example of an HGU-33/P-style flight helmet with the aft communications grommet. This photo clearly identifies Blue Angels "Boss" Gil Rud, who served with the Team in the 1986, 1987 and 1988 show seasons. By the way, 1986 was the first year the Blue Angels transitioned from the A-4 "Skyhawk" to the F/A-18 "Hornet." This example features the more contemporary chrome plated, cast aluminum oxygen mask receivers still in limited use today.

Many times, when you least expect it, information finds its way to you. This is one of those times. Just when I thought I had wrapped up this chapter, Chris Hoglan, a fellow flight gear collector and friend, passed along a contact who reached out to him searching for information on a Blue Angels flight helmet. Max Sabin, Collections Manager at the Fargo Air Museum[5] is preparing a Blue Angels display at the museum and was seeking additional information on a flight helmet donated by CAPT Gil Rud, who was the "Boss" from 1986 - 1988 and led the Team transition from the A-4 "Skyhawk" to the F/A-18 "Hornet." Lucky for me, Max shared some photos of CAPT Rud's flight helmet for inclusion in this book.

[5] https://fargoairmuseum.org/

11
HGU-55/P Flight Helmet

The HGU-55/P flight helmet, made by GENTEX, was introduced in 1982 and saw widespread distribution and use by 1983. To the untrained eye it would be difficult to tell the difference between an HGU-55/P and its predecessor the HGU-33/P.

The top two photos on the left best illustrate the most obvious differences between the 33 and the 55. The top photo is a stripped down PRK-37/P shell the platform from which many U. S. Navy flight helmets are built including the HGU-33/P-style previously used by the Blue Angels.

The middle photo is a stripped down HGU-55/P shell for comparison. The highlighted yellow circles indicate an area of change: The temples (and nape) were cut back on the HGU-55/Ps to facilitate improved comfort and vision for the crew member. When edge rolls and visor housings are attached, a close inspection is required to notice the difference.

Most HGU-55/Ps supplied to the U. S. Navy and U. S. Air Force look similar to the example on the bottom left. GENTEX provides stripped shells to Gibson & Barnes, an authorized distributor, and sole provider of customized flight helmets to the Blue Angels. Gibson & Barnes receives orders from the Blue Angels as new pilots are assimilated into the Team. The Blue Angels provide the helmet size(s) requested (Medium, Large, or Extra Large) as well as the pilots' name or call sign, which is applied on the back of the shell below the crest.

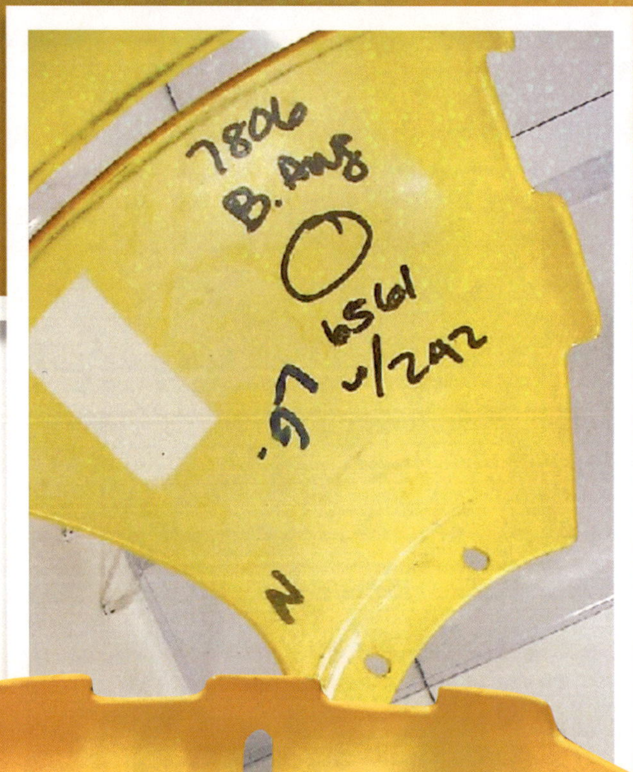

Pictured on the right are photos showing the inside of Blue Angel visor housings and the production marks made by the painter / artists at Gibson & Barnes.

Some of the numbers and markings refer to work orders, visor size, production instructions and aircraft number. The one number common to each of these examples is "6561" which refers to the original DuPont Imron paint code for Blue Angels yellow.

Gibson & Barnes adds an EEK-4A/P single visor kit and gold mirrored visor to the HGU-55/P shell which they designate as the HGU-55/E.

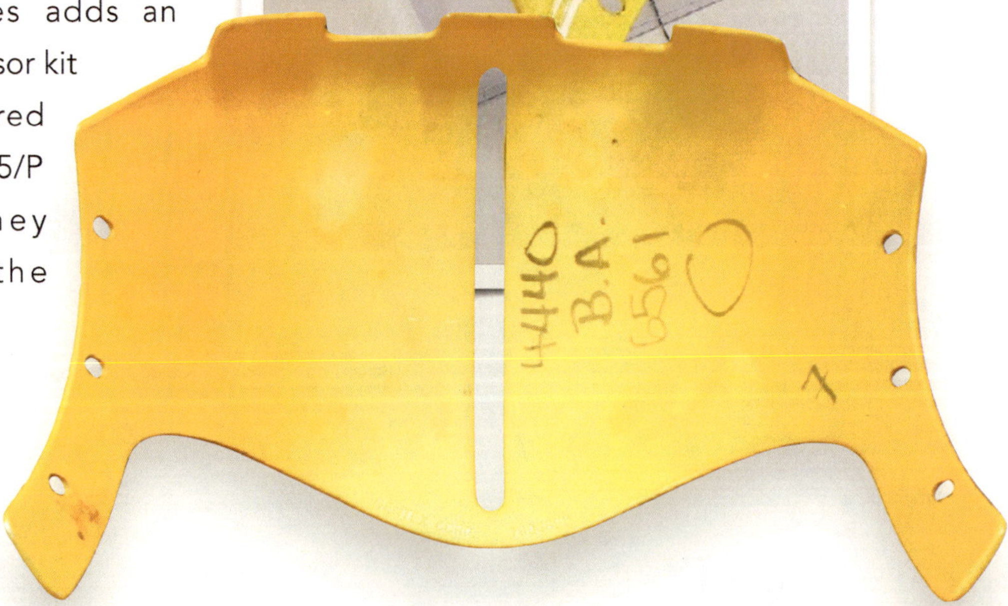

United States Navy Blue Angels Flight Helmets

The HGU-55/E is a Gibson & Barnes designation, not a GENTEX or U. S. Navy designation. A tag is also sewn to the inside liner indicating the date of manufacture, size, order number and customer name. In the example below, NFDS stands for Navy Flight Demonstration Squadron — the Blue Angels.

For reasons that are not clear, the Blue Angels didn't transition to the HGU-55/P helmets until around the mid-1990s, even though fleet pilots were issued and flying with them since the early 1980s.[1] The higher temple trim — resulting in better vision — and lighter weight were cited in the decision to make the switch.

As illustrated with the PRK-37/P shell and the HGU-55/P shell on the first page of this chapter, the differences are slightly more subtle on the completed helmets. Shown below is a PRK-37/P-based helmet on the left, compared to an HGU-55/P-based helmet on the right. Note the difference in the shell trim at the temples and the visor track positioning. Both versions utilize the EEK-4A/P single visor housing assembly.

HGU-33/P "style" helmet built from the PRK-37/P shell.

HGU-55/E (Gibson & Barnes designation) helmet built from the HGU-55/P shell.

[1] James Wegge, Director, Gibson & Barnes

The two images that appear on this and the following page are from the 1996 IMAX production, *The Magic of Flight*[2] (photos from author's personal collection). The image above is of LT Ryan Scholl, who was the Narrator (#7) pilot in 1995, Opposing Solo (#6) in 1996 and Lead Solo (#5) in 1997.

While it's difficult to pick up on the details in these black and white photos, but quality photos of Blue Angels flight helmets with the visors in the up position are difficult to come by, so all the more important to analyze those that are available. This flight helmet — and the one on the next page — are from 1996 and under magnification, appears to be an HGU-55/P based on the position of the visor rails. This would be the

[2] A MacGillivray Freeman Film for the National Museum of Naval Aviation, made possible through the sponsorship of McDonnell Douglas.

earliest examples of the HGU-55/P helmet in use with the Blue Angels — about 13 years after they were introduced to the fleet in 1983.

In the photo below, LCDR Scott Anderson flew the Slot Position (#4) for the 1995 and 1996 show seasons. Even in this inverted position, an examination of the shadowy details suggests this is in fact an HGU-55/P.

The inset photo is a screen capture from a YouTube.com video titled, *Blue Angels Dreams 50th Anniversary Video*. The visor rail "overhang" onto the leather edge roll is further evidence that this is an HGU-55/P.

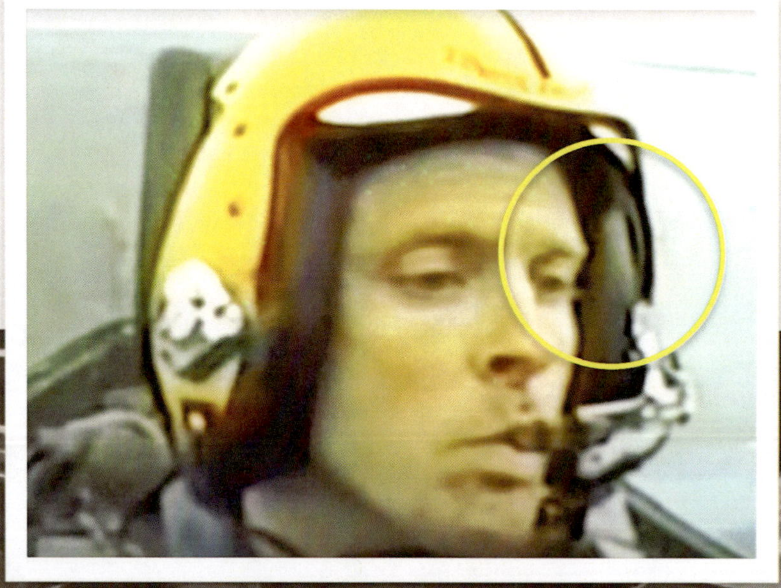

I refused to abandon my search for a former Blue Angels PR who could provide definitive proof of when the Team made the transition, as well as the rationale behind the decision to delay the change. Even James Wegge, Director of Gibson & Barnes was uncertain about these details.

Gibson & Barnes — formerly Flight Suits, Ltd. — was contracted to make custom flight helmets for the Blue Angels since 1984. Mr. Wegge believes the transition to the HGU-55/P didn't occur until after 1993.

The photo on the bottom right belongs to Alan Crane, a fellow flight gear collector from Australia, who commissioned this Blue Angels flight helmet from Flight Suits, Ltd. in 1997. It is clearly an HGU-55/P shell.

In July 2021, during a Blue Angels squadron visit — that included Life Support — I spoke with PR1 Darby Verslius about the late transition from the HGU-33/P to the HGU-55/P. Although this predates her service with the Team, PR1 Verslius mentioned that the Blue Angels have special permission from the Chief of Naval Air Training (CNATRA) to use the HGU-55/P flight helmets, which are no longer authorized for fleet use, until parts are no longer available. Perhaps this was the case with the long-lived HGU-33/P-style helmets? Used until parts were no longer available? Seems like a logical explanation.

This photo is compliments of Major Jeff "D.A.R.Y.L" Mullins, USMC, who flew the Left Wing (#3) position for the 2018 show season and the Slot (#4) position for the 2019 show season. We will learn more about D.A.R.Y.L in Chapter 13, "The Yellow Helmet," and Chapter 15, "Call Signs."

BLUE ANGELS

Although the Blue Angels don't use oxygen masks for flight demonstrations, they don masks for cross-country flights to and from air show sites that typically take them above 10,000 feet, where supplemental oxygen is required.

The oxygen masks shown here are GENTEX MBU-23/P, standard issue for the U. S. Navy. According to the GENTEX Web site[3]. The MBU-23/P is designed for use in tactical aircraft and is supported by NAVAIR 13-1-637-3. The mask is described as lightweight and low profile, providing optimal Positive Pressure Breathing (PPB) sealing capabilities and +600-knot wind blast protection in the

[3] https://shop.gentexcorp.com/gentex-mbu-23-p-oxygen-mask/

(U. S. Navy Blue Angels photos)

event of high-speed or high-altitude ejection.

The two images on this page are of LT Julius "Whiskers" Bratton, Blue Angels #7 (Narrator) for the 2020 show season. These screen shots are from a VIP / Incentive ride LT Bratton was giving to a lucky recipient, visible in the background.

These coveted orientation rides can be punishing from the positive and negative g-forces experienced while the pilot demonstrates the awesome capability of the F/A-18 "Hornet."

It also appears that Lt. Bratton's flight helmet has suffered some "punishment" as well!

The photo on the next page represents a "before" image.

As illustrated earlier in this chapter and in Chapter 10, HGU-33/P Flight Helmet, spotting the difference is difficult. When the visors are in the "down" position, as seen on this page and the previous page, it obscures the details of the trim at the temples, which is a tell-tale indication of the variant. Finding high quality photographs of Blue Angels flight helmets with the visors in the "up" position for research purposes have proven to be challenging. The never-ending search continues.

The mysteries surrounding the Blue Angels HGU-55/P don't end here. There is another curious and elusive anomaly that has defied explanation and is responsible for an inordinate amount of my research time, Integrated chin and nape straps. First, a little background and context.

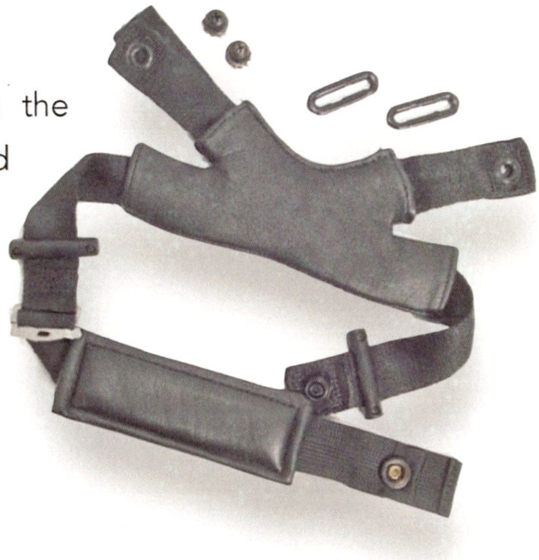

The integrated chin and nape strap (shown on the right) and introduced to the fleet in 1994 replaced the individual chin and nape strap, shown below.

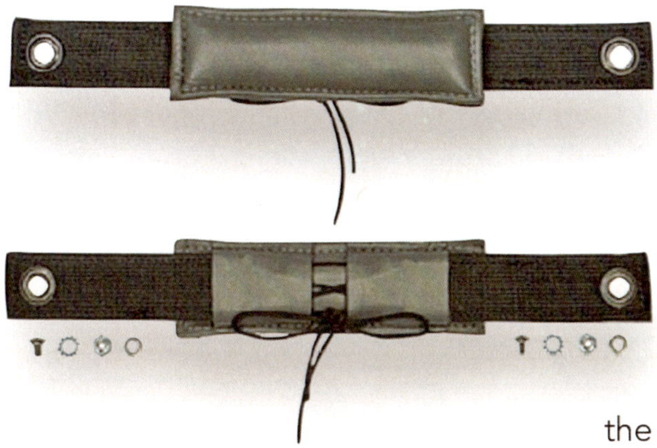

The integrated system allowed for the simultaneous adjustment of the chin and the nape. Tested at 600 knots, the new system purportedly afforded the wearer a more snug, stable and comfortable fit. Apparently not everyone was in agreement. The Blue Angels did not transition to the integrated system and, as of this writing, continue to use the individual chin and nape straps. I have heard some unconfirmed reports that the Blue Angels were not receptive to changing the appearance of their iconic flight helmets, and some felt the individual chin and nape straps were more comfortable.

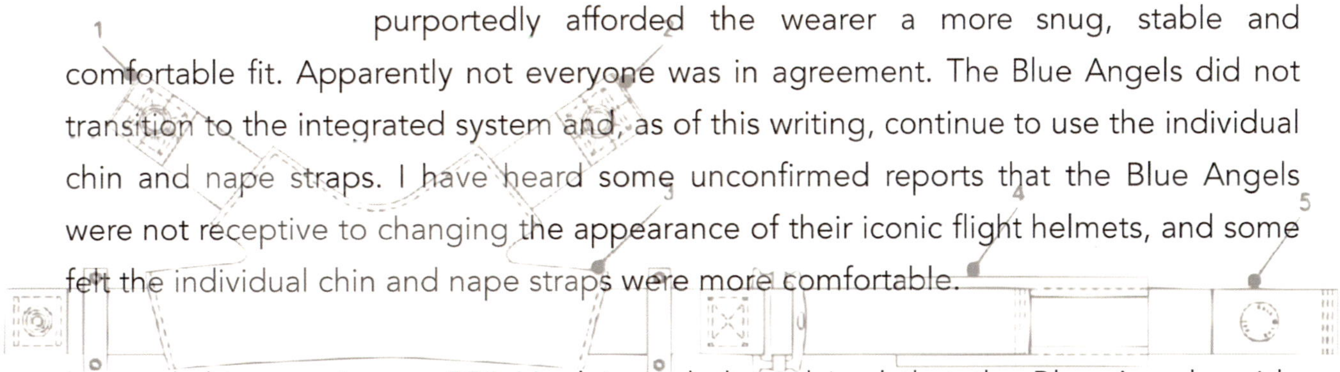

I posed this question to PR1 Versluis and she advised that the Blue Angels pride themselves on tradition and cited the old adage, "If it ain't broke, don't fix it." In addition, she points out that the integrated chin and nape straps have a much wider nape strap / pad which can be uncomfortable for the pilots who need unrestricted head movement during demanding demonstrations.

I posed the burning question regarding the Blue Angels' seemingly delayed transition from the HGU-33/P-style flight helmets to the contemporary HGU-55/P flight helmets. PR1 Versluis explained that the Blue Angels have been granted several equipment deviations by the Chief of Naval Air Training (CNATRA), including their flight helmets. Just like their HGU-33/P-style predecessors, the HGU-55/P has now become obsolete for fleet use but, the Team continues to use them. She explained that CNATRA has authorized the Blue Angels to use a flight helmet until supply has been depleted.

The Blue Angels' sister organization — the USAF Thunderbirds — embraced the idea. The integrated chin and nape straps began appearing on Thunderbirds flight helmets in the early 2000-teens and continue to be in use today.

The interior of the HGU-55/P flight helmet offers a different approach from their HGU-33/P-style predecessors. The latter example used a custom fitted, two-piece, leather-covered liner that was made by Gibson & Barnes / Flight Suits, Ltd. Gibson & Barnes Director James Wegge tells me they ceased making these custom fitted liners when contemporary, commercially available liners in a wide range of sizes became available. This eliminated the requirement for the pilots to be individually custom fitted, a time-consuming process. Instead, the life support shop maintains an inventory of available sizes to be fitted in-house. These liners are easily installed with velcro and can be changed as needs dictate and they are hand washable.

The Blue Angels use the GENTEX Thermo Plastic Liner (TPL), shown below. The preformed TPL consists of a thermoplastic layer assembly (which can be heated and custom-fitted if necessary) and a removable, washable, ventilated cloth cover.

The Blue Angels Flight Helmets on the next four pages are from the 2021 75th Anniversary Blue Angels Team (Author's photos, July 2021).

#1
CDR Brian "G" Kesselring

#2
LCDR James "Red Rocket" Haley

#3
MAJ Frank "CHOMPS" Zastoupil

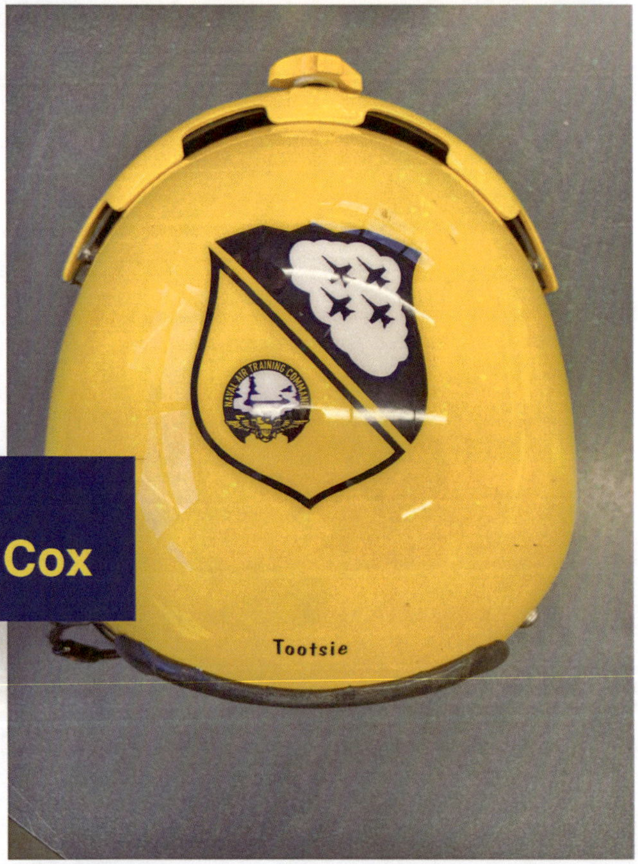

#4
LCDR James "Tootsie" Cox

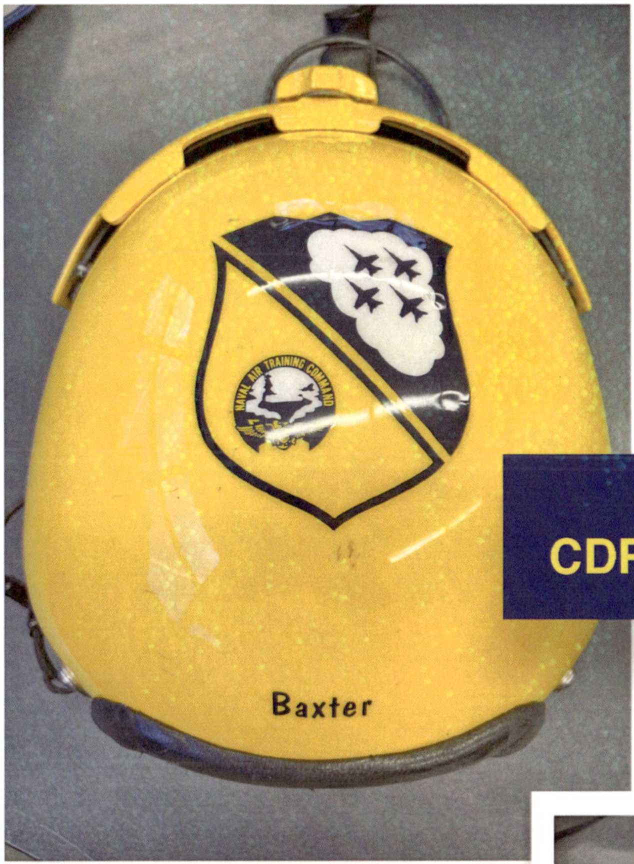

#5
CDR Ben “Baxter” Walborn

#6
LCDR Cary “Chewy” Rickoff

#7
LT Julius "Whiskers" Bratton

#8
LT Katlin "Band Camp" Forster

Even the Blue Angels Supply Officer — called "Suppo" — gets a custom flight helmet complete with call sign. (Author's photo)

On the right is the 2021 flight helmet of the Blue Angels Commander and Leader, CDR Brian Kesselring. Of the countless Blue Angels flight helmets I have studied over the past 25 years, this is the first contemporary example I have seen that did not bear the call sign "BOSS." According to PR1 Travis J. Bradley, Blue Angels Life Support, 2018 - 2020, CDR Kesselring's helmet was initially ordered with the call sign "BOSS." However, CDR Kesselring intervened and requested it be changed to his original squadron call sign, simply, "G." Sorry, I'm not privy to the back story on this one! (Author's photo)

BLUE ANGELS

12
Newbies

Each show season the Blue Angels replace 3 demonstration pilots with 3 new replacement pilots selected from the fleet. This process insures that there are always 3 veteran, second year pilots to balance and train the incoming 3 rookie pilots. The new Team members — known as "newbies" — are selected from a pool of over 80 applicants, all of whom must be carrier qualified, tactical jet pilots with a minimum of 1,250 total flight hours.

The applicants are all volunteers, vying for a coveted position on this extraordinary flight demonstration team. The initial applicant pool is reduced to about 15 candidates to typically fill only 3 demonstration pilot slots, and various other open officer positions, such as C-130 "Fat Albert" pilots, flight surgeon, public affairs, maintenance, etc., as needed.

Early in the process each applicant receives a Blue Angels Guide to "Rushing"[1] which provides additional details, instructions and recommendations. The guide wastes no time and gets right to the point, advising applicants, "We are looking for well rounded, motivated officers to represent the U. S. Navy and Marine Corps 24 hours a day, 7 days a week. Our jobs do not end when we take off the blue suit. As a Blue Angel, everything you say and do is under public scrutiny. Any incident, however minor, can blemish the reputation of the Blue Angels, the Navy and the Marine Corps. If you are not up for that kind of high profile environment, this may not be the job for you."

Applicants are required to obtain their current commanding officer's (CO) approval — in writing — to apply to the Team, and the CO must agree to release the candidate from his or her command if selected.

[1] https://www.blueangels.navy.mil/assets/docs/join/rushing.pdf

Since air shows are the forum in which the Blue Angels execute their mission it is important for candidates to visit show sites as part of their application in order to witness the briefing process and observe how the Team interacts with the community. It is recommended that candidates attend two or three shows between late March and late May of the year in which they are applying to the Team. Candidates are told, "You are on your own when coming to an air show and shouldn't expect any special consideration from show sites because you are an applicant. Do not book a hotel room under the auspices that you are with the Blue Angels."

The guide continues, "When it comes to attending shows, remember, you are not a member of the Blue Angels and thus should not present yourself that way."

"This entire process is a chance for us to get to know you, and for you to get to know the people you could potentially be spending 300+ days a year with. Don't just get to know the person in the position for which you are applying. Make sure you meet every officer on the Team. Don't be shy! We are expecting you to come out of your comfort zone and interact. The bottom line is relax, have fun, be yourself, and enjoy the application process! Remember, we are all fleet personnel and this is just as much of an interview of us as it is of you. Thank you again for your interest in the Blue Angels! We look forward to seeing you on the road!

"Don't be shy! We are expecting you to come out of your comfort zone and interact. The bottom line is relax, have fun, be yourself, and enjoy the application process!"

Blue Angels Guide to "Rushing"

The guide concludes with an outline of what the applicant can expect at a typical air show weekend.

Thursday - There are no applicant functions this day.

The team will arrive Thursday and fly three times (flight from Pensacola, Circles, and a practice demonstration). This is an extremely busy day for us and by the end we are all very worn out and tired. Please don't think we are ignoring you, we just need some time to recoup on Thursday evening.

Friday - Sunday

The brief normally begins at 1400. Applicants are expected to be at the briefing spaces at 1300 when the Team arrives. Socializing outside the briefing space for an hour is informal. We will cycle in and out of the briefing room to visit with applicants during this time. At 1350 the team will enter the briefing room. The brief will begin at 1400. I will come out shortly after 1400 and grab all the applicants and bring them into the room to shake everyone's hand. Once everyone is seated Boss will begin the brief and ask if anyone has a joke to tell to the ready room. Please keep it clean as we often have outsiders in the brief as well. Once the joke is over Boss will conduct the brief which lasts about 10 minutes. Make sure phones are either silenced or off (I recommend not even bringing them in, but that's up to you). During the brief he will go over the exact sequence of the show and the pilots will chair fly at least one maneuver. When the brief is complete he will say "Diamond, Solos, Comm Cart, Tower, Bert, and Narrator, let's have a good one." At this point all the support officers will stand up and shake the pilots' hands, and they will exit the room and conduct the support officer brief. Applicants can either leave the room for the support officer brief or stay in the brief room for a short delta specific brief. Boss will then ask each pilot for their specific goals and the training officer and operations officer will add any additional pertinent information before we break. The solos will depart to a separate brief and the diamond will conduct their own brief. When the solos depart the room all applicants may stay to observe the diamond specific brief and chat with the diamond pilots post brief. After a few minutes of chatting post diamond brief, the applicants should politely excuse themselves so as to provide the pilots with some alone time to get mentally prepared for the demo.

The next function for applicants is the commit.[2] This either happens on Friday or Saturday. This is our opportunity to meet the local public and interact on a more personal level than the crowd line. You are invited to come to most of these

[2] A "commit" is short for commitment and refers to pre-arranged public affairs meetings at hospitals, schools, etc.

events (I will update you each week if that changes). We may, on a limited basis, socialize with you at the commit. However, do not expect it, since we are working. Once the commit is over, we will meet up for drinks. I'll text you all with the information that day. How long to stay is up to you and if you have a flight or reason to depart early; no harm no foul. On Sunday there is just the brief to attend as we depart immediately after the show.

The time outside these windows is your time. You will not receive any special passes, so if you decide to watch the show, you're on your own with regards to getting into the airshow. There is no requirement for you to watch the show so feel free to depart after the brief. There are always questions about the weekend and process, and that's what I'm here for. Here is a rough itinerary based on a normal weekend. Expect this to be sent out each week with specific times and locations.

Rough Itinerary

Friday -
1300 - Arrive at briefing spaces
1400 - Brief begins
1415 - Applicants depart
1900 - Possible commit
2000 - Social

Saturday -
1300 - Arrive at briefing spaces
1400 - Brief begins
1415 - Applicants depart
1900 - Possible commit
2000 - Social

Sunday -
1300 - Arrive at briefing spaces
1400 - Brief begins
1415 - Applicants depart

This initial candidate pool is whittled down, and in early July — about mid-season — the Blue Angels perform an air show along Pensacola Beach. Referred to as "Beach

Week," this is when the candidates who have made the final cut arrive to "rush" the Team in hopes of being selected for those rare, coveted positions.

The first major hurdle is a closed door interview with the 16 squadron leaders. Each officer asks the candidate one question — the same for each candidate — to compare and contrast answers and eliminate variables. In addition to this formal interview candidates are also subjected to informal interviews that take place in social settings to get a sense for who they really are outside of military decorum. Even though the candidates are more relaxed in these social settings (BBQs, picnics, etc.), with other Team members and spouses, they are still subject to scrutiny. One of the simplest, but most important tests to determine is if the candidate will be a good fit for the Blue Angels family.

Regardless of position all candidates are applying for the most important job they will all share, that of public affairs. At the end of the week each new Team member must be selected by the unanimous vote of all 16 officers.

The selection process for the "Boss" follows a completely different protocol, and is an appointment by the Secretary of the Navy. As former Boss George Dom (1997-1998) was quoted, "You can't have the inmates selecting the warden."[3]

"You can't have the inmates selecting the warden."

CAPT George Dom
Blue Angels Boss, 1997-1998

In the 2005 documentary series, *Blue Angels: A Year in The Life,*[4] CDR Russ "Boss" Bartlett (call sign "Beef") welcomes new Team members and tells them, "Your life just changed in ways you can't even comprehend." One of the newbies, Major Dan "Knuckles" Shipley[5], USMC described

---

[3] Podcast by Ryan Northhaft; http://www.youtube.com/c/blueangelphantoms

[4] Henninger Productions, Brian J. Kelly, Director

[5] Currently, Brigadier General

his training and assimilation into the Team as "A lesson of just pure, unadulterated concentration."

"A lesson of pure, unadulterated concentration."

Major Dan "Knuckles" Shipley
Blue Angels, 2004

Without a doubt "Rush Week" is challenging, stressful, and exhausting, emotionally and physically, underscoring the work ethic of each member; however, I am confident the Blue Angels not only work hard, but play hard, as well!

Once the selection process is complete, "Boss" assigns each of the candidates a 5-minute window to call him the following day so he can deliver the news — good or bad. For those not selected, the message is succinct. Boss compliments the candidates performance, and advises them they were not selected. Period.

For those fortunate to make the cut, the delivery is somewhat different. For one successful pilot, LCDR Ted "Bonza" Steelman, Boss — and the other officers — decided to have a little fun at his expense. "Bonza" called Boss at his allotted time slot, and listened intently as he was told, "The competition was exceptionally stiff this year, Bonza. You're a pretty young guy…I thought you did a really solid job, but we're looking for someone a little taller!"

In true professional fashion, Bonza politely responded, "I can totally understand, I mean, I'm only five-ten."

At that moment, the entire Team shouted, "Welcome to the Team, asshole!"

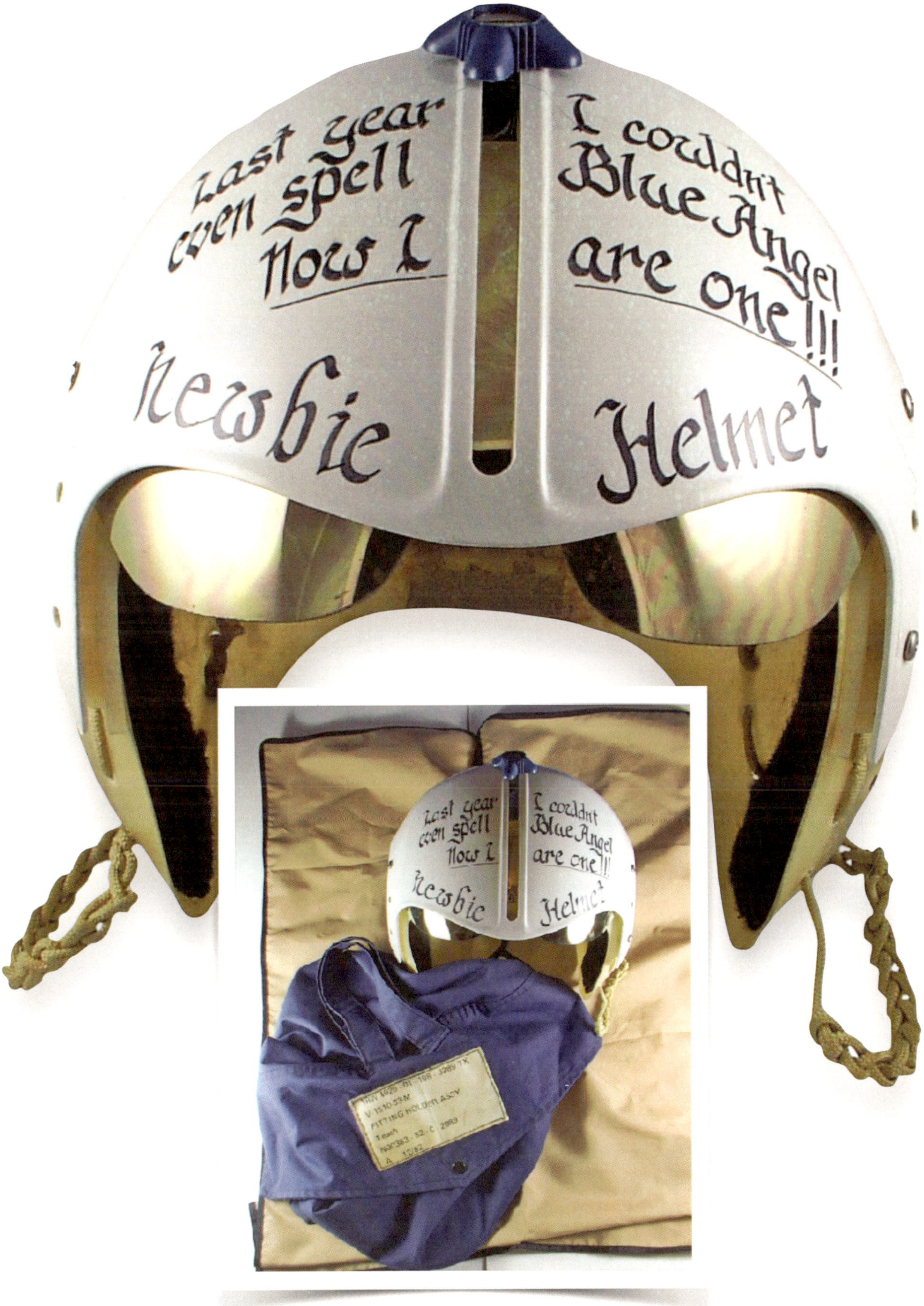
Last year
even spell
Now I
I couldn't
Blue Angel
are one!!!
Newbie
Helmet

The Yellow Helmet

The photos on the previous page are of a most unusual and very unique helmet. As I understand the story this "helmet" was used for various purposes — and a possible skit — during "Rush Week." The "helmet" is actually a shell used to make custom, form fitting "V-TEC liners for flight helmets. The pilot being fitted places a special skull cap on his or her head then dons the shell with the V-TEC liner inside, while a life support technician pours a two-part foam product through a hole at the crown of the shell. The foam expands inside the V-TEC liner and forms a precise contour of the wearer's head. After the foam dries and hardens, it is removed, trimmed and covered in leather. The final product is fitted into the pilot's helmet for a custom fit. It is technically known as V-TEC TRUE CONTOUR FIT INTERLINER FITTING HOLDER.

The crude "helmet" is fitted with a visor housing and classic Blue Angels gold mirrored visor for that special touch. Perfect for a little light hazing and harmless humiliation.

In the 2005 documentary series, *"Blue Angels: A Year in The Life,"* once the new members have been selected and begin their assimilation into the Team, they wear their khaki uniform until the final air show of the season in November. Only then are they allowed to wear the iconic Blue Angels flight suit — called "the Blue Suit." Since they are not yet flying with the Team, and to set them apart from those aviators currently in the demonstration, their suit name tags are void of their position number. Once the Team members complete their training, and the "newbies" fill the slots, only then do they get their position numbers embroidered on their name tags. Some "newbie" tags are shown below (author's private collection).

Unlike their U. S. Air Force counterparts — the Thunderbirds — the Blue Angel pilots are required to wear a "blue suit" while flying a blue jet, with very few exceptions. The Thunderbirds use standard, USAF-issued, green CWU-27/P flight suits for home-based show practices (Nellis AFB / Indian Springs).

This chapter would not be complete without mentioning an iconic Pensacola watering hole, McGuire's Irish Pub. Their Web site offers the following description, "McGuire's Irish Pub, first opened in 1977 as a small neighborhood pub in a shopping center. In 1982 McGuire's moved to its current location, Pensacola's original 1927 Old Firehouse. Inside the pub you'll find a turn-of-the-century, New York Irish Saloon themed 615-seat restaurant."

In 2017, McGuire's decided to pay tribute to the Blue Angels by displaying some Team artifacts obtained through generous donations. Once word spread of the project, current and former Team members began offering flight helmets, flight suits, gloves and photos to show their own appreciation for McGuires, a Pensacola institution. The artifacts are now proudly displayed in two dining rooms and has spilled out into a hallway, making this one of the most significant Blue Angel collections in private hands.

The love for the Blue Angels doesn't stop here. Until a few years ago the Blue Angels officers — a total of 16 — would hunker down in the squadron's briefing room following rush week to deliberate and select the final candidates for next year's team. Successful candidates are required to receive a unanimous vote from all 16 officers in order to be selected. As you can Imagine getting 16 people to agree on *anything* is a challenge! As a result, the deliberations can get spirited, and often run well into the early hours of the next day. They stay the course until they unanimously agree on who will be selected.

In recent years, McGuires has offered to host the deliberations. The officers enjoy a change of venue, are provided with a private dining room so as not to be disrupted, and afforded an opportunity to enjoy a nice meal. Understandably, the date and time of this meeting is not public information and the proceedings are private, for obvious reasons. Oh, to be a fly on the wall!

By the way, McGuires Assistant General Manager is Perry Hunter, the son of Blue Angels pilot Herb Hunter, who flew in the 1957, 1958 and 1959 show seasons.

A final note on "Newbies." Due to the Corona Virus / COVID-19 pandemic of 2020, the Blue Angels were forced to cancel their entire show season. As a result, the Team decided against making any changes to the numbered demonstration pilots (1 through 6) for the 2021 show season...with one exception: LCDR Brandon Hempler. As he had already flown an unprecedented 4 seasons with the Team, it was time for a change. LCDR Hempler will be replaced by CDR Ben "Baxter" Walborn, who previously flew with the Blue Angels as a solo pilot from 2009 to 2011.

Best of luck transitioning to the F/A-18 "Super Hornets" and the 75th anniversary of the United States Navy Blue Angels!

13

The Yellow Helmet

This brings us to The Yellow Helmet.

The Yellow Helmet emblazoned with the coveted Blue Angels crest is without a doubt a highly coveted icon that only a few pilots have earned the right to wear. Unlike flight suits and flight gloves, pilots receive only one flight helmet[1] and when they depart the Team, it is presented to them in a display case to serve as a constant reminder of their service as one of the best-of-the-best.

"The Yellow Helmet. Icons are powerful tools. I think this helmet really represents a culture."

CDR John "Gucci" Foley
U. S. Navy Blue Angels
1990 - 1992

In the Introduction CDR John "Gucci" Foley was quoted, "The Yellow Helmet. Icons are powerful tools. I think this helmet really represents a culture."

A deep rooted culture indeed. A culture that only grows stronger with the passing of every show season. So coveted are these yellow helmets that very few find their way into private collections.

[1] Rare exceptions would be replacement for severe damage, ejection, or pilots returning to the Team following several years between tours.

Rightfully so the vast majority remain with the pilots they were issued to and handed down to family members. A small number are donated to aviation museums. Sadly, others are repainted, taped over or worst, lost or stolen. My point is that with so few of these helmets produced — as compared to the number of helmets produced for the rest of the Navy fleet pilots — it is a challenge to photograph and examine authentic specimens. Patience and commitment are key.

U. S. Navy Blue Angels photo.

As we know, the Blue Angels helmets have not always been yellow. In 1946 the Team wore cloth flight helmets that were standard issue to fleet pilots during this period. Even after the introduction of painted hard shell helmets, introduced in 1948, the Blue Angels maintained their H-1 flight helmets in U. S. Navy, standard-issue white paint. It wasn't until the fall of 1951 that we see the Blue Angels debut their yellow helmets, a 70-year tradition that is not likely to change.[2]

In the 1993 documentary, *U. S. Navy Blue Angels: Around the World at the Speed of Sound,* "Gucci" said, "My Dad was in the Army, and he took me to an air show...I saw some jets flying and that sparked something in my heart. As it turned out, I went to the Naval Academy and started flying jets off of carriers, but the Blue Angels were always that spark...as it turned out, the Blue Angels were at that air show."

Fortunately, through the assistance of many friends and contacts, I have been able to convince a couple of current and former Blue Angels pilots to share some personal, human interest stories about what their Yellow Helmet means to them. Here's what they had to say.

[2] The gold visors were added in 1970 and have remained in service through today.

LCDR Rick "Timber" Young
Narrator #7 - 1993
Opposing Solo #6 - 1994
Lead Solo #5 - 1995

In his official Blue Angels biography, we learn Rick Young hails from Burlington, Vermont and graduated from Rham High School in Hebron, Connecticut where he played football, baseball and varsity tennis. He attended the University of Connecticut, graduating in 1983, earning a bachelor of science degree in business.

Following graduation, Rick attended Officer Candidate School at NAS Pensacola where he was commissioned an Ensign in February 1984. He completed primary flight training at NAS Whiting Field, Florida and transferred to NAS Kingsville, Texas for intermediate and advanced jet training. Rick earned his wings of gold in September 1985.

Rick joined the Blue Angels in October 1992 and served as the Narrator for the 1993 show season, the Opposing Solo (#6) for the 1994 show season and the Lead Solo (#5) for the 1995 show season. When he joined the Blue Angels, Rick had accumulated more than 3,200 flight hours and 450 carrier landings. His decorations include the Navy Commendation Medal, Navy Achievement Medal and various personal and unit awards.

With a little investigative work — and a lot of luck — I was able to locate Rick with the assistance of a mutual friend. In an initial email exchange, Rick expressed his eagerness to participate, and told me, "I am glad to help." He added, "It's your Blue Angels...The team doesn't survive without exceptional followers and fans. We are a team full of humble people representing the extraordinary men and women on the front lines in the Navy and Marine Corps."

"I am glad to help. It's your Blue Angels...the team doesn't survive without exceptional followers and fans."

LCDR Rick Young

LCDR Young, like all Blue Angels pilots who served before and after him, is passionate about his time as a Team member. He tells me that his interest in aviation began at an early age, first fueled by his father's service in the United States Air Force, Air National Guard and 33-year career with Pratt & Whitney Aircraft Engines.

This exposure to aviation influenced not only LCDR Young but, two of his bothers as well. His younger brother followed in his footsteps and became a naval aviator while his older brother earned his wings as a U. S. Army helicopter pilot. LCDR Young said he attributed his desire to fly the result of his dad's influence, adding, "What more can sons ask for?"

LCDR Young's father also impressed upon his son's the importance of a college degree. As a senior in high school, he knew he was college bound. He graduated from the University of Connecticut, while his younger brother was accepted into the United States Naval Academy. Although these two paths were very different, LCDR Young told me, "During that time, I was able to experience the Navy through his [younger brother's] eyes and from the outside looking in." He continued, "That was the spark that interested me in the Navy. I learned that the Navy was not just boats. I grew to understand the critical need for the 'tip of the spear air power' that the Navy could project. I could possibly be a part of that." And, that's exactly what he did.

"I did not get selected the first time around."

LCDR Rick "Timber" Young

LCDR Young shared that he was familiar with the Blue Angels, having witnessed them perform on occasion at air shows. His path crossed the Blue Angels many times when he flew the air show circuit as an F-14 "Tomcat" demonstration pilot.

LCDR Young said, "The first time I applied to the Blues, I was an instructor pilot and Landing Signal Officer (LSO) at the training squadron VF-124.[3] I did not get selected the first time around." This is not unusual.

[3] Fighter Squadron 124 (VF-124) "Gunfighters" (disbanded 30 September 1994) was a U. S. Navy fleet replacement squadron (FRS) located at NAS Miramar, California.

He continued, "After the first bite of the apple and seeing what the Blues were all about, I knew I needed to rededicate myself to applying again." The year that followed, before he was eligible to reapply, LCDR Young said he was invited to be a Commander Air Group (CAG) Landing Signal Officer[4] (LSO) aboard the *USS Kitty Hawk*. "It was a great opportunity to work for one of the best CAGs in the Navy at the time."

"I sat down with CAG during our job interview and I was completely honest with him regarding my intentions to reapply to the Blues." Without hesitation, CAG agreed to support him and allow him transfer if he got selected! LCDR Young's response was, "WOW. A win win. Work for an astonishing person, keep pilots safe landing aboard the aircraft carrier, fly F-14s and still get to chase my dream. I was blessed to have that opportunity."

I inquired about the Blue Angels application and interview process, rush week and selection to the Team. LCDR Young said, "To start with, the Team is able to pick its own replacements, both officer and enlisted, with the exception of the Boss. He / she is picked by a board composed of ex-Blues and flag rank officers.[5]"

"During the show season, [candidates] travel to as many [air] shows as you can while working your fleet job. You get to know the officers and enlisted. More importantly, they get to know you."

---

[4] Landing Signal Officers are naval aviators specially trained to assist and facilitate the safe recovery of aircraft landing aboard aircraft carriers through two-way radio communications with the pilots.

[5] Rear Admiral (O-7) through Admiral (O-10).

"Near the end of the application process, the Team narrows their selections down to a small group. That group is invited to come down to Pensacola to spend additional personal time with the Team and their families." "After that weekend, you wait for the call. A 'nerve wracking' time to say the least. You stay focused doing your fleet job all the while watching the phone."

"Then, you get the call from the Team, one way or another. Always very professional, either way."

LCDR Young said he received his phone call from Boss while the other officers listened in on the conversation. He described it as "One of the longest 3-minute conversations in my life!" He continued, "During the call, your mind is running warp speed without ever taking a breath. When the words of acceptance finally come — maybe after a ribbing or two — it feels like a dreams come true, and the weight of the world has been lifted. Then, the pent up anticipation begins."

When asked about assimilation into the Team, LCDR Young told me, "After 75 years in existence, believe me the Team has it down to a science. From selection, to a "newbie" in khakis to a Team member in a blue flight suit, it is a well orchestrated evolution. Crawl before you walk, walk before you run and expect to test your limits."

"I have never — and will never be a part of something again so unique in my life. To be selected to such an historic and small group of people and to represent the men and women on the 'tip of the spear' in the Navy and Marine Corps is something that not one of us took for granted. The Navy is full of individuals that have the drive and talent that we had and could have taken our place in a heartbeat. Being in the right place at the right time is not just a slogan. I thank God for the chance and hope I didn't slander the precious opportunity."

Echoing these sentiments is Capt. Greg Wooldridge, USN (Retired), who served as Blue Angels Boss for the 1991-1992, 1993 and 1996 show seasons. In the documentary, *"U. S. Navy Blue Angels: Around the World at the Speed of Sound,"* Boss Wooldridge was quoted: "I believe every Naval aviator aspires to achieve all those tenets that you need to have to become a Blue Angel." Very inspiring words, indeed.

I asked LCDR Young what the Blue Angels' iconic yellow helmet and gold visor symbolize and mean to him. His response was a rare introspective into the soul of a Blue Angel pilot: "As I look at the helmet on my shelf, I notice with the visor down and the brilliant yellow paint that it reflects light and images from all angles. You can't escape it. Try taking a picture of it without getting unwanted lights and images in your shoot."

"It represents just that. It is iconic and inescapable when stared upon. It reflects the images back of those who gaze at its uniqueness. I sit here thinking how lucky I was to be given an opportunity to wear such a helmet. To be an average guy given a rare chance to represent the extraordinary. What can anyone ask for?"

In the documentary, *"U. S. Navy Blue Angels: Around the World at the Speed of Sound,* LCDR Young's teammate, LT Doug "Dino" Thompson, who served as #3 (Left Wing) and #4 (Slot) for the 1992-1994 show seasons, was quoted: "I'm just a normal, plain old fleet aviator that got lucky…lucky enough to make it through flight school and into carrier aviation and get picked up by the Blue Angels." Humble men, aspiring for perfection.

In closing, LCDR Young said, "There, by the grace of God, I stood in an unforgiving, tight-ass flight suit, saying, 'Please don't let me screw this up or let my brothers down.'"

I hope you will allow me to say, Well done, Sir. Thank you for your service as a Blue Angel and to our great nation.

Major Jeff "D.A.R.Y.L." Mullins, USMC
Left Wing, #3 — Slot Pilot, #4
2018 - 2019

According the the Blue Angels' Web site, Major Jeff Mullins is a native of Memphis, Tennessee, and graduated from Saint Benedict at Auburndale High School in 2004, where he lettered in wrestling and track. He attended Embry Riddle Aeronautical University, and graduated with a Bachelor of Science in Professional Aeronautics in 2006. Jeff earned his commission as a Second Lieutenant in the United States Marine Corps through the Platoon Leader's Course in 2006, and reported to The Basic School at Marine Corps Base (MCB) Quantico, Virginia, to complete training.

Jeff joined the Blue Angels in September 2017 and flew the #3 Left Wing position for the 2018 show season and the #4 Slot position for the 2019 show season. He is a 2015 graduate of the U. S. Navy Fighter

(U. S. Navy photos.)

Weapons School (TOPGUN), accumulated more than 2,200 flight hours (over 500 combat hours), and earned numerous service awards.

I caught up with Maj. Mullins in July of 2021 in Pensacola, Florida and he was kind enough to sit down with me and share his personal thoughts on being a Blue Angel pilot. Coincidentally, we met at a coffee shop on Palafox Street, just a few blocks from where the famous (or infamous) Trader Jon's was formerly located. This seemed fitting.

Blue Angels flight suit name tag of Major Jeff Mullins, USMC, #4 Slot position, 2019 show season.

(Author's photo and private collection)

I asked Maj. Mullins to share his earliest memories of aviation and key influencers while growing up and his desire to become a pilot. Maj. Mullins tells me that he was "born into a family of pilots,"adding, "My dad is a pilot, and all four of my brothers are pilots." Maj. Mullins said growing up in a flying family, he knew that aviation would be part of his life from an early age, and the early exposure helped guide his decision to fly for the military. I was impressed to learn that Maj. Mullins learned to skydive at the tender age of 11 and had about 4,000 jumps before he flew his first aircraft!

The desire to fly for the military was confirmed when the events of 9/11 stunned the U.S. and the world. Maj. Mullins was only in high school but, felt compelled to make applications to the U. S. Air Force Academy, the U. S. Naval Academy, as well as USAF and USN Reserve Officer Training Corps (ROTC). He landed a USAF ROTC scholarship at Middle Tennessee State University; however, there was no guarantee of a flight slot. He learned through friend that the U. S. Marine Corps offered guaranteed flight contracts. He applied to, and was accepted his freshman year to Embry Riddle Aeronautical University, and the USMC Officer Selection Office accepted him into the Officer Candidates School (OCS), where he would earn his commission as a second lieutenant.

Prior to "rushing" the Blue Angels, Maj. Mullins was the Pilot Training Officer with VMFA(AW)-533[6], MAG-31[7] aboard Marine Corps Air Station (MCAS) Beaufort, South Carolina, flying the Boeing F/A-18 "Hornet" which would prove to be a valuable asset. Maj. Mullins deployed twice — once to MCAS Iwakuni, Japan, and once to Bahrain, in support of Operation Inherent Resolve.[8] Rounding up an already impressive resumé, Maj. Mullins is also a TOPGUN graduate.

Before considering application to the Blue Angels, Maj. Mullins' had been exposed to the Team at airshows and as a student pilot at NAS Pensacola. He shared, "I had a fellow squadron pilot selected to the Blues when I was a new guy in the fleet. Up until that point, I had ruled out ever applying, because I didn't think it would be possible to get selected." Maj. Mullins describes his squadron mate's selection as a "wake-up call" and, "If you don't apply for the things you want to do, even if there's a slim chance you will get it, then the chance of ever getting it are zero."

[6] Marine All-Weather Fighter Attack Squadron 533.

[7] Marine Aircraft Group 31.

[8] Operation Inherent Resolve is the U. S. military operation for the international military intervention against ISIL in Iraq and Syria, and a closely-related campaign in Libya.

I asked Maj. Mullins to share his thoughts and experience around his application to the Blue Angels, the selection process and "rush week."

He said, "There is a formal and informal process associated with applying to the Blues. You submit an application, and then attend 2 or 3 airshows." Observing the pilots while "on-the-job" gives the candidates a taste for what they are about to embark upon. Maj. Mullins continued, "At the airshows, you attend the brief as well as the Blues' scheduled social commitments. The informal part is going out for an hour or so each night and socializing with the Team so you can get to know them and they get to know you as a person and not just an applicant."

"In early June, the Team narrows it down [applicants] to around 6 pilots and 2 or 3 other 'finalists' for each [open] position.[9] The finalists travel to NAP Pensacola during the week of the Pensacola Beach Airshow.[10] That week, everyone does a formal interview. Every officer on the Team is a part of the interview. [Candidates] also attend a brief and debrief on Wednesday of that week. The pilots also conduct a strength test that simulates the forces the spring imparts on the flight controls. And, there is an informal part of the interview that involves attending a dinner with the Team and their families."

Maj. Mullins added, "During this week, all finalists get measured for the uniform items they will be issued so that they are ready when whoever gets selected shows up [to officially join the Team] in September. [Candidates] also fill out an official biography and take a photo in front of a jet [Blue Angels] that they will use when the Team announces the new members." Maj. Mullins admitted, "It is a little bit of a surreal photo to take because you are smiling for a picture that you don't know will ever be seen."

Maj. Mullins said on the Saturday morning following "Rush Week" and prior to the Pensacola Beach Airshow that afternoon, each candidate is given a very narrow and specific "call window" to phone the Boss' office to learn of their fate. The call is short

---

[9] This could be any open officer positions, such as the Maintenance Officer, Public Affairs Officer, Flight Surgeon, etc.

[10] Also referred to as "Rush Week."

and to the point — either you made the Team, or you didn't. He describes the anticipation of this call as an "emotional rollercoaster" because "it is a huge determination of the next 2 to 3 years of your life as well as a dream most who are in that position have been chasing."

Maj. Mullins told me he was not selected the first time he rushed the Team, which is not unusual. There are documented cases of naval aviators and USMC pilots rushing the Blues 3 and 4 times before being selected. A true testament to the drive, determination and dedication these pilots possess.

> **"My Helmet Fell Apart"**
>
> **Following an airshow in Cleveland, Ohio, and upon completion of our shutdown procedures, I removed my flight helmet – per protocol – and it fell apart!**
>
> **There are 3 screws on either side of the visor housing securing it to the helmet. I guess that over time, these screws loosened up, and as I proceeded to take my helmet off, the entire visor assembly fell off! It sort of disintegrated! Now, there's FOD all over the cockpit!**
>
> Maj. Jeff Mullins, USMC

I was curious about the challenges of assimilating onto the Team. Maj. Mullins assured me that "The Team is very welcoming and you are a part of the Team from day one showing up. You're training starts that day. It is in the Team's best interest to welcome you, get you up to speed, and start the training as soon as possible for the next show season."

I was interested in what the yellow helmet means to Maj. Mullins. He said, "My biggest responsibilities with the Blue Angels — besides flying — was to inspire a culture of excellence and service to country through flight demonstrations and community outreach." This is the Blue Angels mission in a nutshell.

He continued, "The blue flight suit and yellow helmet embody this mission statement. People immediately recognize the blue flight suit and yellow helmet. When you remove these iconic items, no one knows who you are. Individual pilots are faceless members of a larger organization. Everything is bigger than the individual — it's all about the organization."

CDR Russ Bartlett — Blue Angels "Boss" for the 2003 and 2004 show seasons — echoed Maj. Mullins' sentiment in the 2005 documentary, *Blue Angels: A Year in The Life.*[11] CDR Bartlett — call sign "Beef" — was giving his new Team members a motivational talk and was quoted, "It's not *you* they want to talk to, it's the suit. They don't care who Russ Bartlett is, and they won't remember a year from now who Russ Bartlett is, they'll just remember that they met Blue Angel 1, 2, 5, 6, 7…that's all. It's that simple, boys and girls." This certainly is a gut-check in humility.

I am grateful for Maj. Mullins taking time to meet with me and share his inspirational story. Best of luck to you, Sir, with whatever the next chapter of your life has to offer, personally and professionally. I am confident you will not only succeed, but excel.

[11] Henninger Productions, Brian J. Kelly, Director.

14
VIP Flight Helmets

In an ongoing effort of self-promotion, the Blue Angels regularly give incentive flights or VIP rides to key community influencers, members of local media and celebrities. The desired outcome is to encourage those chosen for these highly sought after flights to provide positive feedback to the community about the United States Navy and promote recruiting. It's all about publicity.

The fortunate chosen few are outfitted with dark blue "VIP" flight suits and matching flight helmets. The VIP flight helmets are essentially the same as those used by the pilots — HGU-55/Ps — but the paint and graphics colors are reversed. The pilots wear a yellow helmet with dark blue graphics, and the VIP helmets are dark blue with yellow graphics. An example of each are shown below for "compare and contrast" purposes.

I was privileged to be invited to visit the Blue Angels Life Support Shop on several occasions starting in the mid-1990s and into 2020. The VIP helmets — like their yellow counterparts — typically have a name on the back below the crest and just above the edge roll. The exception would be spare helmets. The Leader traditionally uses "Boss" and the other pilots use a call sign or first name. Call signs will be covered in detail in Chapter 13.

The two images on this page are Blue Angels flight helmet graphics, or decals as there are commonly referred. The one on the left is dark blue and white and used on the yellow helmets worn by the pilots on the flight demonstration team, typically #1 through #8

The decal on the right is a yellow / gold and white and intended for use on the dark blue VIP flight helmets used for those lucky passengers selected for incentive flights.

(Courtesy of the John Slezak collection)

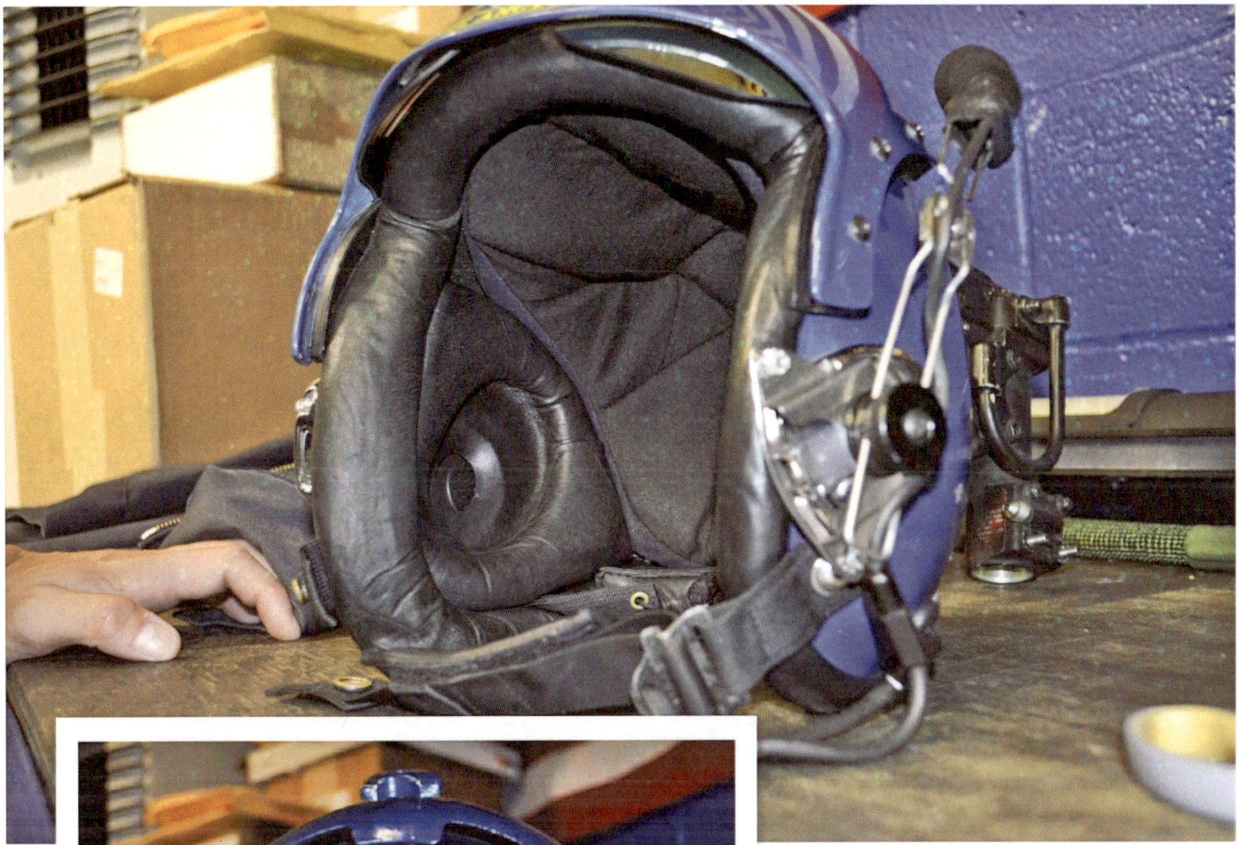

These images (author's photos) were taken inside the Blue Angels life support shop while attending the Homecoming Air Show in 2018. My host was PR1 Travis J. Bradley, whose hand can be seen in the top photo.

One of my goals on this visit was to examine and photograph a VIP flight helmet. This example was clearly an HGU-55/P as evidenced by the dramatic trim at the temples and the position of the visor housing rails, which "encroach" on top of the leather edge roll.

This one is simply identified as "VIP / PA" but I learned that some were given special call signs. The photo above is Dan Schaffner, an ABC Action News Anchor, as he prepares for an incentive ride at McDill AFB for AirFest 2010 (USAF photo). Although the flight helmet is slightly out of focus, the call sign is still readable: Crybaby! One of my all-time favorites!

A very unique "VIP" is the Hollywood actor Ernest Borgnine, who is one of only a handful of people who hold the coveted title of "Honorary Flight Leader." Other notable title holders include Bob Hope and Jon "Trader Jon" Weissman (see chapter 19). Mr. Borgnine is wearing a DH-151 flight helmet with Hardman oxygen mask receivers. This undated photo is from the book, *Reflections of Blue - A Pictorial History of the U. S. Navy Blue Angels,* by Carol Knotts.

Here is another comparison of old vs new. The top photo is of actor Harrison Ford — a pilot himself — as he prepares for his ride of a lifetime in the back seat of an F/A-18 "Hornet." Mr. Ford sports an HGU-55/P flight helmet. The inset photo is of reporter Art Harris in the back seat of an A-4 "Skyhawk" prior to a VIP incentive ride, circa 1977. Mr. Harris is wearing a DH-151 flight helmet with Hardman Engineering & Tool Company oxygen mask receivers.

I have been researching Blue Angels flight helmets for the better part of 25 years and an avid collector of military flight gear for almost 40 years. In that time, I have never seen a Blue Angels VIP flight helmet in any collection, public or private. Only a handful have ever been produced, which tends to explain their scarcity. But, there's no shortage of photos. Here are a few.

PO2 Travis Simpson, Crew Chief for Blue Angels #7, assists Dan Schaffer, ABC Action News Anchor prior to his "Key Influencer Ride" at McDill AFB in 2010. (USAF photo)

VIP/PA

Martha Sugalski of ABC News affiliate WFTV, Orlando, Florida, prepares to take flight with Blue Angels Narrator (#7) LT Julius "Whiskers" Bratton (U.S. Navy Blue Angels photo).

The Yellow Helmet

On this page are a few photos that I was able to capture while visiting the Blue Angels Life Support Shop in the Summer of 2021. They only maintain two (2) VIP flight helmets, a large and a medium.

United States Navy Blue Angels Flight Helmets

15
Call Signs

Who doesn't remember the 1986 Paramount Pictures blockbuster hit *Top Gun?* Whether you're a fan or a critic, one thing is certain — the U. S. Navy got a lot of milage out of this film, and moviegoers became fascinated with fighter pilot call signs.

The history and origin of call signs are varied. They are used extensively in the aviation community and are usually associated or inspired by a humorous play on the aviator's name, prominent personality traits, or exploits during his or her flying career.

"Negative, Ghost Rider, the pattern is full."

"Top Gun"
Paramount Pictures
1986

Prior to entering into the age of political correctness, there were few, if any rules governing call sign protocol. The results were always funny, often embarrassing, and occasionally crude, to say the least.

By way of the latter example, I was granted permission to share the story of a Vietnam era USMC F-8 "Crusader" pilot's call sign, provided he could remain anonymous. As I am told, he was known as "MACE" which is an acronym for "Mexican Ace." In military pilot vernacular, an ACE is a pilot who successfully shoots down 5 enemy aircraft. Of course, "MACE" was quick to clarify that he didn't shoot down 5 Mexican aircraft. However, he did claim to "service" 5 Mexican "working girls" in one night at a "palace of ill repute" in San Luis, Mexico (just across the border from Marine Corps Air Station Yuma, Arizona) along with his plane captain.

As he walked out to the flight line the following Monday morning, there were 5 Mexican flag decals mounted on the rear ventral fins of his trusty "Crusader!" When his Commanding Officer saw his newly awarded "ACE" status, he almost got thrown in "hack" for "deliberately damaging U. S. Government Property!" That's his story, and he's sticking to it!

These stories were not uncommon at a time when testosterone flowed as freely as the beer, and egos were tougher than a two-dollar steak. Over time, society began to develop a growing intolerance for behavior that some referred to as "boys-being-boys," particularly when the combat exclusion for women was lifted and the fighter pilot community became "co-ed."

However, not everyone was so quick to get on-board with the culture change. Amid law suits over "offensive" call signs in the past 10 years, John "Toad" Lehman, former Secretary of the Navy (1981 - 1987), and A-6 "Intruder" Bombardier / Navigator, argued that political correctness had doused the profession's spirit.

Secretary Lehman was quoted in a 2011 article in *Proceedings Magazine,* saying, "Those attributes of naval aviators — willingness to take intelligent calculated risk, self-confidence, even a certain swagger — that are invaluable in wartime are the very ones that make them particularly vulnerable in today's zero-tolerance Navy."

Others have charged that the chain of command is not willing to identify a systemic problem in the aviation community regarding inappropriate call signs.

In response, Lehman said, "I've been in and around naval aviation for like 40 years and I've never heard of a call sign that had a nasty turn to it." He went on to say, "The notion that call sign guidelines need updating 'is a crock.'"[1]

Regardless of your position in what has become a hot debate, one thing is for certain, call signs are not going away.

---

[1] http://content.time.com/time/nation/article/0,8599,2011189,00.html

On the lighter side, U. S. Navy LT Kara "Hulk" Hultgreen[2] — the first female carrier-based fighter pilot in the U.S. Navy — received her call sign because of her ability to bench-press 200 pounds, her 6-foot frame and a play on her last name. Following a television appearance in which she wore significant makeup, her squadron mates rewarded her with the revised call sign "Revlon[3]."

While these call signs seem innocent enough, some aviators have taken issue with what they have described as racial harassment. In a 2019 article written for military.com, two African-American pilots filed formal complaints alleging racial bias.

A U. S. Navy investigation, led by the Commander of Naval Air Forces, cited inappropriate conduct by instructor pilots for failing to treat the students "with appropriate dignity and respect" through the use of discriminatory call signs. The Chief of Naval Air Training was directed to formalize a call sign assignment and review process, to include appropriate peer board representation for minority and female aviators. In addition, squadron training was directed on "unconscious bias and stereotype threat."

Many applauded the decision, validating pilots' concerns — for years — that the Navy needed to address their treatment and allegations of racial bias.

This case stemmed from a series of black student aviators who were offended by their "assigned" call signs, including "Radio," which was a reference to a movie that portrayed a mentally challenged black man. Other students reported receiving call signs such as "8 Ball" and "Kazaam," the latter being a character played by Shaquille O'Neal. Other minority pilots provided evidence of race influenced call signs, such as, "MC," short for "MC Hammer," and "Snoop,"a reference to "poor rapping abilities."

The investigation also cited unfair criticism and public shaming of some students by their instructor pilots through social media. A specific case involved a student who

[2] Killed on 25 October 1994 while on approach to the USS *Abraham Lincoln* during a routine training mission off the coast of San Diego.

[3] Cosmetics Company.

was publicly shamed by his instructor, claiming, "That dude is so bad. I can't even comprehend how someone does not have the most basic level of self-awareness and realize he's going to kill himself or someone else."

Another aviator was singled out when he misinterpreted the instructor's order and blew the canopy off an F/A-18. A class T-shirt was printed to exploit the mishap bearing the slogan, "Once you pop, the fun don't stop."

Needless to say, political correctness is playing an ever increasing role in our society and the military is no exception. However, not everyone shares this sentiment. A USMC Major — who will remain anonymous — was quoted, "That's the price to pay in this community. You do stupid shit, you get made fun of...or you kill yourself or others" Speaking of the dismissed students, the major said, "His attrition was his small price to pay. Frankly, we did his ass — and the community's — a favor. But this bullshit makes me professionally embarrassed."

Regarding the Command investigation, one of the complainants was quoted, "...I have serious concerns about the content of the investigation report and the manner in which the investigation was conducted. I find it troubling that the only conclusions from this review is that call signs need to be changed. There are larger challenges facing minorities and women in Naval Aviation than just being given demeaning and inappropriate call signs."

A Naval Air Forces spokesperson said, "Treating all service members with dignity and respect is something the Navy takes extremely seriously and if an EO (equal opportunity) complaint is filed, it is appropriately investigated and, if applicable, appropriate action is taken. As Sailors we represent the Navy 24/7. Our core values of honor, courage and commitment apply to all aspects of our life, including online conduct. Any form of harassment, bullying, hazing, stalking, discrimination, retaliation, or any other type of misconduct that undermines dignity and respect [is] not consistent with our core values."

DEPARTMENT OF THE NAVY
CHIEF OF NAVAL AIR TRAINING
250 LEXINGTON BLVD SUITE 102
CORPUS CHRISTI TX 78419-5041

1610
Memo 00/ 408
25 Jul 19

MEMORANDUM

From: Chief of Naval Air Training

Subj: CHIEF OF NAVAL AIR TRAINING (CNATRA) CALL SIGNS POLICY

1. Purpose. This memorandum sets forth procedures for Chief of Naval Air Training (CNATRA) and Fleet Replenishment Squadron (FRS) commands on assigning call signs to Naval Aviators (Pilots and Naval Flight Officers).

2. History. Call signs have a rich naval heritage and are a time-honored tradition in Naval Aviation. Carriers, air wings, and squadrons all have defined call signs and it is only natural that individual aircrew would as well. They have proven valuable as another level of authentication/identification in making a clear distinction between aircraft in the air, particularly during times of intense, fast-paced operations. They provide brevity in communication among the tightly knit fabric of naval aviation that squadrons represent. Originally required for tactical communication during operations where radio transmissions could be intercepted, call signs prevented individuals from outside of the naval unit from identifying the speaker. Call signs remain an important symbolic representation of the warrior culture that exists throughout Naval Aviation.

3. Today. By linking a proud history with modern Naval Aviation, call signs create a sense of camaraderie and belonging, and should enhance morale and promote esprit-de-corps among ready room personnel. However, if a call sign is inappropriately assigned, it becomes a distraction to the command and can deny the individual receiving the call sign their rightful sense of belonging in the squadron.

4. Action. Each member of the Navy is entitled to be treated with dignity and respect, and to work in an environment free of harassment and unlawful discrimination. Our call signs must be in keeping with this directive and continue to promote a sense of camaraderie and belonging among all Naval Aviators. Accordingly, call signs shall not be lewd, graphic, or sexual in nature, nor should they be assigned on the basis of race, ethnicity, national origin, sex, sexual orientation, gender identity, or religion.

5. Procedure. Call signs may be assigned through a formal or informal process, including call sign review boards, kangaroo courts, or detachment student reviews. However, multiple members of the command shall assemble to review and recommend all call signs. To the extent practicable, the call sign process should include as diverse a membership as possible. The Commanding Officer (CO) shall provide oversight and shall approve all call sign assignments in writing. The CO may delegate this authority to the Executive Officer.

Subj: CHIEF OF NAVAL AIR TRAINING (CNATRA) CALL SIGNS POLICY

a. Call signs can be instructive or play off a first name or surname. Good humor, creativity, and cleverness is encouraged, but call signs shall not bring shame or discredit to the Navy or to the individual receiving the call sign, nor shall they be offensive, malicious, or hostile.

b. In the call sign process, the CO must be made aware of any secondary meanings or pop culture references. Call sign review members shall advise the CO of such secondary meanings or pop culture references even if they are unintentional to the call sign assignment.

c. All Navy personnel shall be cognizant of the changing nature of public perception when determining the appropriateness or continuing appropriateness of a call sign.

d. When a call sign is issued, the individual receiving the call sign shall be advised of the rationale behind the assignment. If an individual believes their call sign is not in accordance with the directives outlined in paragraph four, they are strongly encouraged to inform their CO of the issue to provide the command an opportunity to address their concern.

6. Leadership. Leaders are responsible for setting an appropriate command environment and culture. We will guide our personnel to the right behavior and demand change when required. We must be sensitive to those situations in which a call sign is not in alignment with our ethos because it humiliates, demeans, belittles or intimidates someone within our community. This is our ready room, in our fleet; Naval Aviators are expected to take ownership of this issue and address any inappropriate conduct immediately. What is appropriate for a call sign will evolve, and we must continue to recognize and embrace those changes.

G. N. HARRIS

Releasability and distribution: This notice is cleared for public release and is available electronically only via Chief of Naval Air Training Website, https://cnatra.navy.mil/pubs-instructions.asp

2

The letter on the preceding pages from the Department of the Navy, Chief of Naval Air Training, details the outcome of the investigation and guidance on call signs policy.

The directives in this memorandum are detailed and leave little room for interpretation. The following highlighted paragraphs set the tone for the future of assigning U. S. Naval aviator call signs:

"However, if a call sign is inappropriately assigned, it becomes a distraction to the command and can deny the individual receiving the call sign their rightful sense of belonging in the squadron."

"Accordingly, call signs shall not be lewd, graphic, or sexual in nature, nor should they be assigned on the basis of race, ethnicity, national origin, sex, sexual orientation, gender identity, or religion."

"Call signs can be instructive or play off a first name or surname. Good humor, creativity, and cleverness is encouraged, but call signs shall not bring shame or discredit to the Navy or to the individual receiving the call sign, nor shall they be offensive, malicious, or hostile."

"When a call sign is issued, the individual receiving the call sign shall be advised of the rationale behind the assignment. If an individual believes their call sign is not in accordance with the directives in paragraph four, they are strongly encouraged to inform their CO of the issue to provide the command an opportunity to address their concern."

Whether you agree or disagree with the findings and directives of this investigation, one thing is certain, this will have a direct impact on a long-standing tradition that will forever change the culture of the fighter pilot community.

So, what about the Blue Angels? They are no different from their brothers and sisters in the fleet — they all have call signs.

It is unclear when the Blue Angels began including their call signs on the back of their flight helmets, but following a visit to the archives of the National Museum of Naval Aviation at NAS Pensacola, we know the practice began as early as 1954.

The photo on this page is the back of a Blue Angels GENTEX H-1 flight helmet that belonged to Captain Chuck Hiett, the first U. S. Marine Corps pilot to join the Team. His call sign — Jar-Head — is clearly displayed on the rear of his helmet. This time-honored tradition continues today.

Convincing a pilot to reveal how he or she was awarded their call sign — especially if it is particularly embarrassing or sensitive — is a challenging task. It typically requires you to have met the person, developed a bond, trust or friendship and then — and only then — can the process of coaxing the "back story" begin. Oh, and the big reveal usually occurs at a bar...with you picking up the tab!

I've been on this earth for over 6 decades, and have only had the opportunity on a handful of rare occasions to learn the details behind a pilot's call sign. This is without a doubt a tough nut to crack. Despite my best efforts, I failed miserably to convince very many current or former Blue Angels pilots to "fess up" on how they received their call signs. Not all is lost. I have managed to scour that vast information repository we call the Internet for some examples of Blue Angels call signs that were posted — or leaked — into the public domain.

Blue Angels Call Signs

Boss

Let's start with the easy ones. Regardless of who the Commander / Leader of the Blue Angels is from season to season, they all arrive with their own, personal call signs picked up in flight school, while in the fleet or serving on a combat tour. However, while serving on the Team, a long-standing tradition dictates his new — albeit temporary — call sign will be "Boss." By way of example, Russ "Beef" Bartlett was the Blue Angels commanding officer for the 2003 and 2004 show seasons. During this two-year assignment, he was known exclusively as "Boss."

Doc

Similar to the "Boss," the squadron flight surgeon is typically assigned the call sign "Doc."

MO

Continuing on what seems like a theme, the Blue Angels' Maintenance Officer is predictably and unceremoniously given the call sign...you guessed it, MO!

Fudge

Arguably one of the most notable Blue Angels call signs was the #2 Right Wing pilot for the 1992 and 1993 show seasons, Larry "Fudge" Packer. Seriously? Evidently, I'm not the only one who finds this moniker as brutally crude. In the 1993 A&E television documentary, *Blue Angels — Around the World at the Speed of Sound"*[4], each of the six demonstration pilots are introduced by the program narrator using the pilot's name and call sign, and often referred to throughout the 96-minute program by only their call signs...except for one: "Fudge." Production oversight, or soft censorship? You make the call.

A similar story appears on the Web site "BlackFive" where a contributor posted, "Back in the 90's, the Blue Angels had a LT Larry Packer as one of their demo pilots. Of course, his call-sign was 'Fudge.' But I doubt he was gay. Anyway, so when the Blue Angels Narrator gave the names of all six pilots (Aviators) as they did their walk, he would give everyone's name and call-sign except for poor LT Packer. LOL! Those crazy Squids!"[5]

Poptart

Major Katie "Poptart" Higgins-Cook, USMCR was selected by the Blue Angels in the summer of 2014 to be the C-130 "Fat Albert" pilot for the 2015 show season. Her selection marked the first female pilot in the Blue Angels' history.

---

[4] Produced by Rob Stone.

[5] https://www.blackfive.net/main/2010/08/derogatory-call-signs-fagmeister-goes-hot.html

History aside, we are here to talk about call signs. According to Maj. Higgins, the call sign was acquired innocently enough, the result of a drunken gorge on pop tarts. However, while flying a mission over Afghanistan in support of Operation Enduring Freedom, a British Joint Terminal Attack Controller (JTAC) inquired — over the radio — how she got her call sign. Before she could reply, her co-pilot responded, "It was her stripper name before she found Jesus!" This sealed the deal on "Poptart!"

No

In a rare departure from the "rubber stamp" assignment of "Doc," the 2006 Blue Angels Flight Surgeon — a female — Tarah Johnson, was given the call sign "No" as in, No Johnson. Wow!

Lawman

Ben "Lawman" Hancock, USMC flew with the Blue Angels for the 1994 and 1995 show seasons. "Lawman" received his call sign because he served as a Trooper with the Arizona Highway Patrol prior to his U. S. Navy service and selection to the Team. This particular back story is one near and dear to me, since I also served as a law enforcement officer. Thank you for your service, Sir — times two!

Gucci

LCDR John "Gucci" Foley flew with the Blue Angels for the 1990, 1991 and 1992 show seasons as the narrator, opposing solo and lead solo, respectively. During a public radio interview,[6] Lcdr. Foley was asked the inevitable question, "How did you get your call sign?" He said that in the early 1980s, while training to be a fighter pilot, a group of pilots decide to go out to a bar at the end of a training day, and agreed to meet up in the lobby. As they gathered, he noticed that everyone was in jeans and t-shirts, while

6 99.1 WPLR, New Haven, Connecticut.

he was sporting a black button down shirt with a black leather tie! He said, "That wasn't even cool in the 80s!" The rest of the pilots asked, "What's that? That's Gucci!" LCDR Foley added, "Well, no fighter pilot wants to be called Gucci! You want to be 'Hitman,' 'Viper,' 'Iceman,' 'Maverick!' And, so, it stuck!"

Soupy

Major Bill "Soupy" Campbell, USMC served in the #8 position, the Blue Angels' Events Coordinator for the 1980 and 1981 show seasons. Does anyone need any help with this one?

D.A.R.Y.L.

I was fortunate to make contact with Major Jeff Mullins, USMC who flew the #3, left wing position for the 2018 show season and the #4, slot position for the 2019 show season. With a little persuasion, I was able to convince him to share the "back story" on how he got his call sign, D.A.R.Y.L.

According to Maj. Mullins, the call sign is based off of the 1985 Paramount Pictures[7] science fiction film, *D.A.R.Y.L.* The film is about a military experiment in artificial intelligence, which intended to produce a "super soldier." D.A.R.Y.L. is an acronym for Data-Analyzing Robot Youth Life-form.

Maj. Mullins tells me, "I was given the call sign while in my first squadron by our Canadian exchange pilot. It is a play on the fact that I was younger than my peers, since I graduated college and commissioned in the Marine Corps when I was 19. I also tended to come off as a bit emotionless due to my dry sense of humor, and the poker face I would keep when getting messed with as a new pilot in the squadron."

---

[7] Paramount Pictures (USA & Canada), Columbia Pictures (International.

IROC

In a 2014 interview conducted by the United States Naval Academy Alumni Association & Foundation, LT Mark Tedrow reportedly earned his call sign — "IROC" — when he arrived at his first U. S. Navy squadron driving a Camaro IROC Z-28.

CHOMPS

An aviation podcast called "The Pilot Base" conducted an interview of Major Frank Zastoupil that aired 01 February 2021. The podcast was was titled "Being a Blue Angel Pilot" and, as you might have guessed, the subject of call signs was addressed.

According to Maj. Zastoupil, his call sign — CHOMPS — came from an overseas deployment. He was an AV-8B "Harrier" pilot on the *USS Iwo Jima* on a Marine Expeditionary Unit headed to the Red Sea. As he prepared for a 2:00 a.m. launch — the first real mission of the deployment — his fuel gauge and engine digital gauge drops to zero. He said, "I knew I couldn't launch." With no choice but to abort his mission before launching, his squadron mates thought an appropriate call sign would be CHOMPS: Choked On The Mission Possibility!

BRUTAL!

Turkey

In an interview with Ryan Nothhaft[8], LT Jerry "Turkey" Tucker says, "Call signs are not chosen, they are awarded." LT Tucker recounts his days with the Reserve Air Group (RAG) when his class would go out on tactical training flights, also referred to as dog

---

[8] blueangelphantoms.com

fights. During that time, he was hoping he would get the call sign "Friar," a play on "Friar Tuck." He said, "It seemed to fit." Upon returning from a tactical training flight, he drew a a picture of a frying pan with what he described as a turkey inside the pan, wings extended, shooting down an aircraft. A squadron mate took one look at the picture and said, "That's not a fryer, frying chicken, that's a turkey!" LT Tucker lamented, "And, it stuck!"

I have always been and will continue to be fascinated by call signs and their colorful "back stories." This practice is a time-honored tradition, steeped in history and will likely continue to thrive despite the myriad of rules and regulations spawned by an oversensitive and politically correct society. Just my biased opinion.

It is also said within some circles that in cases of call signs derived from outrageously embarrassing or taboo circumstances, there may be two versions of how the call sign was awarded, depending on the audience. Only those close enough — and trusted — to the pilot or squadron may be privy to the details of the "real" back story. Others outside this inner-most circle, such as media and casual acquaintances, would get a believable, albeit watered-down, "scrubbed" version so as not to offend the person making the inquiry, without risking unnecessary exposure or embarrassment. Although this may seem like just another "old wives' tale," I have good reason to believe this has been, and will continue to be a way to quietly circumvent a system ripe with over regulation. Again, just my opinion.

"Maverick? Did your mother not like you or something?"

Kelly McGillis as "Charlie"
Paramount Pictures, TOP GUN - 1986

16
The Ground Crew

The Blue Angels' enlisted maintenance and support team is comprised of approximately 120 Sailors and Marines with alternating crews of about 45 team members traveling to each show site. All career-oriented enlisted Sailor or Marine applicants come recommended for Blue Angel duty by their current commanding officer. Applicants go through extensive screening, including interviewing with the members of each of the 15 squadron work centers. The keen selection process secures the squadron's tradition of excellence, ensuring the Blue Angels are a direct reflection of the professionalism of today's Sailors and Marines. After completing their Blue Angels tour, individuals return to the fleet to continue their naval careers.

Selected enlisted personnel volunteer for a three-year tour with the squadron. Though every team member brings skills in a distinct job specialty, each is expected to work beyond that specialty, contributing to the overall effectiveness of the Blue Angels. The squadron consists of seven distinct departments, jointly responsible for guaranteeing command readiness. A tribute to this dedicated team is the fact that the Blue Angels have never canceled an air show due to a maintenance problem.[1]

One of these teams — Aircrew Survival Equipmentman — is a specialty rating with a concentration on maintaining the survival gear used by U. S. Navy and United States Marine Corps flight crews. Formerly known as Parachute Riggers[2], they are responsible for inspecting, maintaining, and

---

1 https://www.blueangels.navy.mil/team/enlisted.htm

2 Parachute Riggers (PRs) were reclassified as Aircrew Survival Equipmentman in 1965.

repairing parachutes, search and rescue equipment, survival kits, medical kits, flight clothing, flight helmets, protective wear, night vision equipment, aircrew oxygen systems, liquid oxygen converters, anti-exposure suits, and g-suits.

All U. S. Navy and Marine Corps flight squadrons employ Aircrew Survival Equipmentmen, and the Blue Angels are no exception.

Coincidentally, Aircrew Survival Equipmentman class "A" technical school is conducted at Pensacola NAS. Following this 12 week course of study, most Aircrew Survival Equipmentman are assigned to squadrons, aircraft carriers or other Navy ships carrying aircraft, or Naval Air Stations. There they learn more advanced skills in their rating through on-the-job training.

Because this job is responsible for the life safety of flight crews, PR students must maintain a 90% passing grade or better to complete the training, making it one of the most challenging courses at the Naval Aviation Technical Training Center.

On the previous page is an example of the Aircrew Survival Equipmentman (PR) badge as it would appear on the sleeves of sailors with the ranks from Petty Officer 3rd Class (E-4) to Petty Officer 1st Class (E-6).

Bob Moore
Blue Angels Aviation Structural Mechanic (AM)
1952 - 1953

Another team — Aviation Structural Mechanic rate, or AM — is responsible for maintaining a host of aircraft airframes and structural components, including, but not limited to fuselages, movable control surfaces, empennages, hydraulics, oxygen systems, egress systems, landing gear, and launch and arresting systems. In addition, duties include minor repairs to aircraft skin, welding and painting.

In the summer of 2019, I was fortunate to make contact with Bob Moore, a Blue Angels AM3 in 1952 and 1953. This is his story.

Bob is a native of Galveston, Texas and knew early in his life that he wanted to serve his nation in an aviation capacity. He first tried to join the U. S. Air Force but, was told there was a 9 month processing period. Anxious to serve, he enlisted in the U. S. Navy in April of 1951. Bob was 19 years old.

One short year after enlisting, Bob had orders to report to the U. S. Navy Blue Angels at NAS Corpus Christi, Texas. Bob said in the early 1950s, enlisted personnel weren't applying for and competing for open positions on the Blue Angels — you were assigned...just like any other duty assignment. Bob said he never heard of the Blue Angels prior to this assignment!

The Blue Angels were rebuilding following their reassignment to combat status for the Korean War. Butch Voris was selected as the commander — his second with the Blue Angels. Bob said when he arrived at NAS Corpus Christi, Butch Voris was the first person to greet him. He described Voris as a "Nice guy."

"(Butch) Voris was a Nice Guy."

Bob Moore
Aviation Structural Mechanic
U. S.Navy Blue Angels
1952 - 1953

One of Bob's responsibilities was painting the Team's show aircraft — the F9F-5 "Panther" — and, the pilots' flight helmets. According to Bob, they used the same lacquer paint for the aircraft and the flight helmets. Bob described painting 2 different style helmets in his 2 years with the Team: He described one as "ribbed" and the other as "smooth." Bob said the "ribbed" helmets were like the ones used in the 1954 action / drama movie, *The Bridges at Toko-Ri*. Unfortunately, Bob couldn't remember the

specific helmet nomenclature, but the "smooth" one was likely the H-1, and the "ribbed" one could have been the H-2, H-3 or H-4. Bob recalls serving under CAPT Ray Hawkins, who replaced Butch Voris, and we know from historical photos that CAPT Hawkins used an H-2 helmet.

The very rare photo on this page is of Bob Moore painting Zeke Cormier's H-1 flight helmet. In February of 1954, CDR R. L. "Zeke" Cormier relieved Ray Hawkins as the Officer in Charge, or "Boss."

When I spoke with Bob, he told me the Blue Angels didn't have a paint room — they painted wherever it was convenient! After the paint was sprayed and dry, he applied the U. S. Navy wings and Blue Angel "banner" decals. He remembers painting about 6 flight helmets during his tour. Unfortunately, he has very limited photos from his time on the Team due to personal cameras being forbidden on base.

I can't help but notice how times have changed, namely, the total lack of personal protective equipment. In todays environment state and federal occupational safety and health administration (OSHA) regulations would mandate a dedicated paint room or booth to protect the environment, as well as the use of a proper respirator, eye protection, coveralls and gloves. I can't help but wonder how many times Bob painted his watch!

After Bob left the Blue Angels, he reported to VF-111 "Sundowners" before separating from the Navy. In civilian life Bob retired after a career with the Federal Aviation Administration as an Airframe & Powerplant mechanic.

Thank you for your service, Bob. And, thank you for sharing your stories.

An unidentified Life Support specialist sands an APH-6 flight helmet in preparation for a new, yellow Blue Angels paint job. This photo is courtesy of Richard Keane, who served as the #3 Crew Chief from 1969 to 1971. According to Keane, the Blue Angels flight helmets were painted in-house during this time period, just prior to Protection, Inc. taking over the customization duties.

(Photo credit: Ron Renfro)

PR2 Robert Fugate meticulously polishes the gold mirrored visor that has become an icon of the Blue Angels flight helmets since 1970. It was imperative that these gold visors be free of any scratches that may cause a distraction to the pilots while flying in very close formations.
(Photo credit: U. S. Navy Blue Angels)

PR2 Robert Fugate
Blue Angels Life Support
1985 — 1988

In late Spring of 2021, I was contacted by Robert Fugate, an Aircrew Survival Equipmentman Second Class[3], who served with the Blue Angels from 1985 to 1986 — the last 2 years of the McDonnell Douglas A-4 "Skyhawk" — as well as from 1987 to 1988 — the first 2 years of the Boeing F/A-18 "Hornet." PR2 Fugate responded to inquiries I made via several social media platforms searching for subject matter experts on Blue Angels life support. I rolled the dice, and "Mother Luck" smiled favorably upon me.

My initial contact with PR2 Fugate was via email, where I learned of his considerable flight helmet experience with the Blue Angels, including a one-week temporary duty assignment (TDY) with Flight Suits, Ltd.[4] in El Cajon, California. This was pure gold!

Robert hails from Ypsilanti, Michigan, entered the U. S. Navy in 1981 at Recruit Training Command, Great Lakes, Illinois, and was selected to join the Blue Angels in November of 1984. He retired from the Naval Reserve as a PR1 in 2001 after 20 years in the U. S. Navy.

Thank you for your service!

(Photo credit: U. S. Navy Blue Angels.)

[3] Formerly Parachute Rigger Second Class.

[4] Now doing business as Gibson & Barnes.

When Robert joined the Blue Angels in 1984, he said the Team was using the HGU-33/P-style flight helmet, and continued to use it until his departure following the 1988 show season, despite the HGU-55/P being in service to the fleet pilots since 1982 or 1983. The HGU-55/P flight helmets were issued without visor housings (weight saving measure) and used an exposed visor attached to the sides of the helmet shell with bungee straps and snaps. The pilots and PRs wanted to maintain the integrity of the style and look of the Blue Angels helmets as well as insure the gold visors were afforded proper protection, therefore, maintained the HGU-33/P in inventory.[5]

Because these yellow helmets had become so iconic, the PRs took great pride in the manner in which they were handled and presented for public view. Robert explained that the PRs had to learn how to properly hold the helmet so the crowd could be afforded the best view. He says that they had to learn the individual mannerisms of each pilot. As they returned from a demonstration and removed their flight helmets, the pilots would lower their helmets from the cockpit to awaiting PRs on the ground. Some would carefully hand over their helmets while others would just drop it to the PR! Robert said you needed to know what was coming, and make sure you looked good as you recovered the precious equipment!

HGU-33/P-style helmets were also issued to and used by the #7 Crew Chiefs as they traveled to and from show sites with the Narrator (#7) in advance of a scheduled air show. These helmets were painted blue instead of yellow. Also wearing the blue helmets were those lucky people from the sports, music or movie industry who were granted an incentive flight. These helmets didn't include a two-piece, custom fitted, leather covered liner like the pilots received. Instead, the PRs used leather faced, adhesive-backed pads that they could adjust to some degree to make the helmet as well-fitted and comfortable as possible.

Some notable VIPs were presented with a personalized yellow helmet as a cherished keepsake of their incentive ride, typically fitted with a standard-issue tinted visor. A few lucky ones were gifted with the trademark gold visor.

---

[5] An EEK-4A/P visor housing assembly was an available option for the HGU-55/P, as used by the USAF Thunderbirds.

Robert tells me that while in El Centro for winter training in 1986, Master Chief Jack Cragen[6] suggested spending about a week at Flight Suits, Ltd. since it was "just over the hill" in El Cajon, California (about 100 miles to the west). Flight Suits, Ltd. was a GENTEX authorized distributor and the single-source provider of custom flight helmets to the U. S. Navy Blue Angels, as well as the USAF Thunderbirds.

While in El Cajon, Robert was tasked with learning every aspect of pouring a custom two-piece liner, covering it in leather, as well as measuring the pilots for a hand-sewn leather edge roll to increase comfort and reduce blind spots. These details were not taught to fleet PRs as part of the Aircrew Survival Equipmentman training. Robert would share this knowledge with future Blue Angels PRs.

True Confessions

"I was the spare crew chief my last year on the Team (1988), and later became the permanent crew chief for #3, Lcdr. Mark Ziegler. I had to learn each pilot's quirks and necessities should their crew chief be absent for personal or Navy reasons."

"We were in El Centro – affectionately known as "El Sweato" or "El Sucko" – for winter training. I was tasked to be CDR Gil Rud's (Boss) crew chief for that particular day. We performed our early morning turns and released the planes to the pilots. I was rushed due to a mandatory dental appointment that same morning. I launched CDR Rud and the other 3 diamond pilots followed. I completed my dental exam and zipped back. My mindset was, if I launch a certain pilot, I want to recover that same pilot."

"I returned in time to recover CDR Rud with no problem. He came down the ladder and shook my hand as per protocol. He than handed me the flashlight that I left in the cockpit! It hit him in the visor / helmet when he rolled inverted to practice the Double Farvel maneuver! He never chewed my ass out, and I was thankful it happened in winter training where the separation was greater than a normal airshow. Of course, stricter tool inventories were implemented following that incident to insure this would never happen again!"

"I thought I was GONE! The yellow helmet saved CDR Rud and my job!"

PR2 Robert Fugate
Blue Angels Life Support

[6] Served with the Blue Angels from 1986 to 1987.

The photo on the right, courtesy of PR2 Robert Fugate, is a random El Centro "shop photo" of a Blue Angels HGU-33/P-style flight helmet, with an MBU-14/P oxygen mask in use during his service with the Team. (Photo credit: PR2 Robert Fugate, U. S. Navy Blue Angels)

The "gem" on the bottom of this page — courtesy of PR2 Robert Fugate — is a very rare and beautiful example of a "decal" sheet from the A-4 "Skyhawk" era from Flight Suits, Ltd. These decals — silkscreened on clear, adhesive backed acetate — were cut from the sheet and applied to the Blue Angels flight helmets after the yellow paint was applied and prior to the final clear coat (additional details can be found in Chapter 14, "VIP Flight Helmets" and Chapter 19, "Don't Judge a Book By It's Cover").

The Blue Angels "crest" was positioned centered on the rear of the helmet, between the edge roll and the visor housing, and the Blue Angels "scroll" was positioned centered on the visor housing between the forward edge and the visor lock track.

It never ceases to amaze me the many treasures stashed away in personal collections, waiting patiently to be re-discovered. As a collector and historian, I have a keen appreciation for those who share this common passion and a desire to preserve a small piece of history. Without these dedicated efforts, our history would be lost forever.

Thanks to PR1 Robert Fugate for your service to our great nation, and your willingness to share your unique history.

Travis J. Bradley
U. S. Navy Blue Angels
Life Support 2018 - 2020

I first met Travis J. Bradley during the 2018 Blue Angels Homecoming airshow at NAS Pensacola. While there I made arrangements to meet up with long-time friend, Brian DeRitter — a former Blue Angel in life support — and asked him if he could once again assist me with my research. I was hoping to get some reference photos of the Blue Angels VIP flight helmets and, without hesitation, Brian took on the challenge with a smile.

As a former Blue Angel — and always a Blue Angel — Brian had unrestricted access to the squadron. He excused himself and walked through the squadron door like he did so many times 20 years ago. A few minutes later, Brian returned and said, "Follow me." It had been many years since I last entered through those doors. The experience was just as exciting as it was the first time I walked those hallowed halls. It was busy. Very busy. We navigated the narrow hallways filled with the hustle and bustle of two-way pedestrian traffic. After a few turns off the main hallway, we arrived at the life support shop.

Brian introduced me to Travis and provided him with a brief overview of how we met and my life-long obsession with the Blue Angels. Travis had a VIP helmet at the ready and welcomed me to take as many photos as I needed. In between shots, I shared some details about my intention of writing a book on the history of the Blue Angels flight helmets. Travis seemed intrigued, and provided me with his contact details in case there was anything else he could do to assist me. Unfortunately, our time was cut short due to this being a show day, and Travis had work to do in preparation of launching the jets.

Over the next two years, Travis and I remained in touch as he continued to answer my seemingly unending barrage of questions. The time had come to request a formal interview. Travis was receptive to the idea, and after paving the way with Blue Angels Public Affairs, I was granted permission for an on-site interview at NAS Pensacola in September of 2020. This could not have come at a better time, as the Blue Angels had just cancelled their 2020 show season and Travis would be out-processing from the Team effective 01 November 2020.

Blue Angels Life Support
NAS Pensacola

My persistent badgering of Blue Angels Public Affairs was rewarded with an invitation for an on-site visit of the squadron and an interview of Travis Bradley. I made the 11 1/2-hour drive from a previously planned stop in Wilmington, North Carolina to NAS

Pensacola the third week of September, 2020. As luck would have it, I arrived just in time to square off face-to-face with Hurricane "Sally" positioned due south of NAS Pensacola in the Gulf of Mexico and moving on a very slow, westerly track. Having grown up in southwest Louisiana, I was familiar with hurricanes and was more concerned with missing out on an interview appointment than I was about a storm! Besides, this was a minimal hurricane. Well, I was wrong.

Hurricane Sally approached slower than Christmas morning and was churning up the warm Gulf of Mexico waters along the way. As she reached a Category 2 storm[7], she unleashed unrelenting rains causing power outages and widespread flooding throughout the panhandle of Florida and southwestern Alabama. With a forward speed of 5 miles per hour — or less — Sally was quickly becoming an uninvited guest who overstayed her welcome!

I was committed, and felt reasonably secure in my hotel, several miles inland, high and dry on the 4th floor. The evacuation window was now officially closed. By late Tuesday, Sally took a northerly turn and had her sights set on the Alabama / Florida border, with Pensacola on the northeast side of the storm — the quadrant with the most intense wind and rain. Overnight, Sally made landfall, and as is customary with hurricanes, knocked out electrical service along her path of destruction.

The wind and rain pounded the window of my hotel room throughout the night and I finally succumbed to the punishing noise when the power began to fail at about 4:00 a.m. After a cup of coffee to brace for the day ahead, I logged into FaceBook where I learned Travis' home was flooding and he had to evacuate his family to the safety of a friend's home under the worst possible conditions. Priorities changed; safety was paramount; my visit was no longer mission critical.

---

[7] The Saffir-Simpson Hurricane Wind Scale is a 1 to 5 rating based on a hurricane's sustained wind speed. This scale estimates potential property damage. Hurricanes reaching Category 3 and higher are considered major hurricanes because of their potential for significant loss of life and damage. Category 1 and 2 storms are still dangerous, however, and require preventative measures. A Category 2 storm has sustained winds of 96 - 110 mph.

Travis J. Bradley

Travis J. Bradley hails from De Soto, Missouri, and served with the Blue Angels from 2018 to 2020 as a PR1(AW). In layman's terms, these are survival equipment specialists and certified parachute riggers (PR) who oversee critical life saving equipment, such as parachutes, flight helmets, oxygen masks, etc. The "1" represents petty officer first class, and the "AW" is an abbreviation for Aviation Warfare qualified. In a word, impressive.

This interview was conducted via FaceTime, in part due to COVID-19 protocols.

As I would soon learn, impressive would be an understatement to describe this man, and his profession.

Travis said he was inspired by two uncles serving in the U. S. Navy, one is a Master Chief and the other a Petty Officer first class.

PR1 Bradley
Life Support
SN#
E1193657
DANGER DO NOT PULL HANDLE
James Haley

With the interview pre-empted, I hunkered down in my hotel room for 2 days as Sally unleashed her punishing winds and rain upon Alabama and the Florida panhandle. I was thankful for the safety of my hotel room, despite not having electricity or running water. As soon as Sally made her predicted easterly turn, I headed out westbound on I-10 for Louisiana. As I looked at Pensacola in my rearview mirror, my heart grew heavy for Travis, his family and all those affected by Hurricane Sally.

Travis J. Bradley "selfie" prior to his incentive ride.

Fortunately, Travis and his family survived "Sally" and they began the process of returning their lives to some sense of normalcy. Amid the changes was Travis' scheduled separation from the Blue Angels effective 01 November 2020. Due to the storm damage and cleanup efforts following Hurricane Sally, Blue Angels Public Affairs would not permit an on-site squadron visit and interview. However, they did authorize Travis to participate in a virtual interview.[8]

Travis remained eager to participate in the interview and we decided to select a mutually agreeable date following his transfer to his next duty station.

[8] A videotelephony and online chat service through a cloud-based peer-to-peer software platform used for teleconferencing, telecommuting, distance education, and social relations.

Having a large military family, especially Navy, he said this was an opportunity to make something of himself.

Travis enlisted in the U. S. Navy and originally had a plan of doing special operations, but due to undisclosed medical responses, he became a Navy Corpsman.

In 2011, Travis reported to Recruit Training Command, Great Lakes, Illinois for training as a medical corpsman. Upon graduation in 2012, he reported to San Antonio, Texas to continue training as a hospital corpsman. After 6 months, Travis took a leap and requested to cross-rate to an aviation rate, which offered opportunities for him to take part in special programs. The Navy approved his request.

He was granted to the prestigious rate of Parachute Rigger (PR) which not only allows for sailors to work in Aviation Survival, but also provide support for Special Operations, commonly known as the Navy Seals.

With this change in rate, Travis transferred to NAS Pensacola, where he received training to become a Parachute Rigger (PR). While there, he got to witness the Blue Angels every week as they practiced, and recalls saying to himself, "One day, I'm going to be a part of the Blue Angels team!"

Travis mentions that after being in the Navy for about a year, he still had not been out to the fleet. But, that was about to change.

When Travis graduated from NATTC "A" school as a PR, he was selected as one of the first sailors to be assigned to the newly formed VFA-101 "Grim Reapers" which was an F-35C "Lightning II" Fleet Replacement Squadron based at Eglin Air Force Base, Florida.

"One day, I'm going to be a part of the Blue Angels team!"

PR1 Travis J. Bradley

Travis worked hard, and became the first PR to become qualified as an F-35C Plane Captain, and made Petty Officer 3rd Class, citing two mentors at Eglin AFB who made him the sailor he is today: AD1 Scott Garland and Chief Petty Officer Nathan Widener. Prior to arriving at Eglin, both Garland and Widener served as Blue Angels.

In 2014, Travis was transferred to VFA-213 "Black Lions," an F/A-18F "Super Hornet" squadron assigned to the *USS George H. W. Bush*. While there, he deployed to the Northern Arabian Gulf as part of the U.S. operation "Inherent Resolve" which was

intended to counter the growing threat of the Islamic State In Iraq and Syria (ISIS). VFA-213 was the first squadron to drop ordnance on ISIS forces. The ops tempo quickly escalated from shore duty (Eglin AFB) with fifth generation fighters, the F-35C, to a fleet wartime environment in the Middle East and fourth generation aircraft, the F/A-18F. Travis said it was surreal to be part of a squadron dropping live ordnance on an enemy faction. But, there was a sense of pride to be carrying out the mission for which he trained.

Travis — along with VFA-213 — participated in a second deployment as part of operation "Inherent Resolve." The USS George H. W. Bush, Strike Group 8 and VF-213 were the only maritime military asset to be at sea when President Donald Trump was inaugurated. In 99 days, the "Black Lions" deployed 488 air-to-surface munitions, 1,300 rounds of 20mm, and over 550,000 pounds of ordnance on target, boasting a 100% combat sortie rate. Travis described these two combat deployments as the hardest, but most rewarding tours of his career.

At the end of the deployment, the shop lead — a PR1 — got to return home, leaving a vacancy. The skipper — CDR Kevin "K-Robb" Robb — called a meeting at the hangar bay. The Commander called out Travis and said, "I think you are out of uniform, shipmate." Travis' response was, "What do you mean, Sir?" CDR Robb replied, "That crow on your sleeve is missing a chevron. Congratulations, you have just been meritoriously advanced to 2nd class!" With that also came a promotion to shop lead. Well done, Travis!

> **"Congratulations, you have just been meritoriously advanced to 2nd class!"**
>
> CDR Kevin Robb

While deployed, Travis had been working hard building a package for eventual submission of his application to the Blue Angels. Around July of 2017 word came down that the Blue Angels were accepting applications, only one of which would be from the Parachute Rigger rate. One PR to be selected...from the entire U.S. Navy fleet. And, that includes applicants from the United States Marine Corps.

At the time, Travis had applied as a Petty Officer 3rd class, but the Blue Angels were looking for a Petty Officer 2nd class. Travis said that he believes CDR Robb was promoting his application, as he was friends with "Boss," CAPT Ryan Bernacchi, as well as the incoming "Boss," CAPT Eric "Popeye" Doyle, and suspects he informed them of Travis' meritorious promotion.

It was rumored that a deal was struck between Robb, Bernacchi and Doyle, that if Travis was selected for the Blue Angels, they would have to agree to a delayed reporting, as he was needed at VFA-213 for a scheduled U.S. Navy aviation maintenance inspection.

At the end of the day, Travis got the nod, was selected for the Blue Angels and reported directly to NAF El Centro[9], 35 to 40 days after the rest of the team reported. He was concerned he was behind the power curve on his requirements to become "crested." When new Blue Angels team members — "Newbies" — report for duty they are issued the iconic navy blue uniforms with which we are all familiar...except for one difference: The Blue Angels crest is missing. This revered patch must be earned following several weeks of training in squadron history, procedures, teamwork and traditions. Only then is the coveted patch awarded to those who successfully pass this challenging hurdle in a formal cresting ceremony.

The "normal" cresting process is designed to take 90 days. Travis put his nose to the grind stone and successfully completed the requirements in 31 days! These are the type people who make the Blue Angels the best of the best.

---

[9] Blue Angels winter training.

Travis J. Bradley (right) being presented his Blue Angels crest by CAPT (Then CDR) Eric "Popeye" Doyle, AKA "Boss."

I was able to glean a little insight into the "Newbie" pilots as seen through Travis' eyes. During "rush week" or "beach week" (see Chapter 17), there may be about 10 pilots who have made the final cut for perhaps 3 flying positions. During this week, the PRs measure the heads of every pilot candidate so they can get a jump start on what they may need to order for incoming pilots. A similar process is followed for the "Blue Suits" — see Chapter 20.

Unlike the USAF Thunderbirds who assimilate into the Team while wearing their USAF, government-issued, gray HGU-55/P flight helmets. Only when the instructor pilots and Team members agree that the "Newbie" pilots are qualified to fly their assigned position solo are they "awarded" their red helmets — a time honored tradition. The Blue Angels are issued their custom flight helmets immediately after being selected and joining the Team.

Incoming Blue Angels pilots also bring with them their issued oxygen mask from the fleet, which are typically MBU-23/P style, but some holdovers are still using the MBU-16/P style due to comfort, despite the U.S. Navy's best effort to phase them out.

When asked why the Blue Angels still use the individual chin and nape straps and not the integrated chin / nape strap that are common throughout the fleet, he said, "Comfort." The general consensus among the Blue Angels pilots is that the individual straps are more comfortable, so that's what they get! Travis added that he also prefers the custom adjustments the individual nape strap offers.

According to Travis, because of the unique mission of the Blue Angels, the Navy allows them minor deviations — within reason — from strict equipment regulations. A case-in-point is authorizing the flying of a combat jet with a boom mic and no oxygen mask[10] during flight demonstrations and training, and only required to wear their oxygen mask during transit flights due to the higher altitudes.

Above and Beyond

Travis' resume as a sailor are nothing short of exemplary – the reason behind his meritorious promotions.

In addition to his responsibilities as a PR, Travis excelled in his pursuit for excellence. Travis trained for a #7 crew chief, "Chuck" position, (tertiary position), qualifying him for Low Power Turn / Taxi Operator for "Morning Turns." Known as the "7-Geek," it is considered a prestigious position, as this is the only flying crew chief who accompanies the #7 pilot – Narrator – to show sites ahead of the Team to conduct advance preparations prior to an airshow.

Adding to his resumé was the title of Qualification and Certification Ordnance Program Manager, responsible for qualifying other Life Support personnel to perform their job in accordance with the Navy's strict ordnance handling program. This is a job typically held by an officer.

This hard work and dedication resulted in Travis becoming the number one early promote (EP) sailor over all other hand-selected E-5s in the command.

As we wrapped up our virtual interview, Travis concluded by sharing a final proud moment he experienced prior to his separation from the Team on 01 November 2020. In September of that same year, Travis was summoned to the Public Affairs Office by CDR Brian "Boss" Kesselring, who said, "Travis Bradley, you are out of uniform!" You guessed it, Travis received his second meritorious promotion to PR1 and a grade of E-6. Travis will be assigned to shore duty at NAS Pensacola to become a military instructor for new PR sailors and marines assigned to the fleet.

Congratulations, PR1 Travis J. Bradley and thank you for your service to our great nation.

---

[10] The Blue Angels routinely use oxygen masks on cross-country flights to and from air show sites as well as prolonged flights above 10,000 feet.

Crew Chiefs

The Naval Aircrewman is a U. S. Navy rating awarded to qualified, enlisted flight crewmembers. The rating is commonly referred to as crew chief, and he or she is responsible for maintaining the safety and airworthiness of the aircraft for which they are assigned.

It is a common understanding that although pilots fly the jets crew chiefs own them. Few, if any pilots, dispute this. The crew chiefs are entrusted to perform all pre-flight duties — typically performed by the pilot — prior to all flights. They are trained to start and perform engine run-ups, and are taxi qualified.

According to the Blue Angels Web site, "The Crew Chief work center is composed of Navy and Marine Corps maintenance technicians in various aviation specialties. Two crew chiefs are assigned to each aircraft, and they are responsible for performing engine turn-up operations and flight controls / instruments checks, as well as daily turn-around and preflight inspections of the aircraft. Each crew chief goes through extensive training to get a working knowledge of the aircraft and its systems."[11]

It should come as no surprise that the duties and responsibilities of the crew chiefs are critical to the safety of the pilot, aircraft and mission, whether it's a combat sortie or an aerial demonstration. Personnel who earn these coveted positions are awarded a qualification badge — commonly referred to as wings — intended to recognize their unique accomplishments. These wings are worn with pride.

[11] https://www.blueangels.navy.mil/team/enlisted.htm

The Yellow Helmet

In the fleet, enlisted personnel working on aircraft carriers wear impact resistant, flight deck crewman's helmets to afford the wearer some degree of head, eye and ear protection. These helmets are color-coordinated — along with their jerseys and life jackets — that identify their unique job responsibilities on the flight deck.

The various roles are categorized by seven colors: Purple, yellow, blue, brown, green, red and white.

In the case of the Blue Angels, The entire ground crew team wear blue for uniformity. However, crew chiefs are further set apart from their peers on the Team with their own, custom flight helmets, similar to those worn by the demonstration pilots.

It is not clear when Blue Angel flight helmets were first issued to crew chiefs, but we know of one example from the F-4 Phantom era that resides in the archives of the National Museum of Naval Aviation at NAS Pensacola.

On the following page is an example from the F-4 era (1969 - 1974), this APH-6D, dual visor helmet belonged to Petty Officer First Class (PO1C) Gerald Harvey. According to Jared Galloway, Museum Technician (Aeronautics) at the National Museum of Naval Aviation, PO1C Harvey was the crew chief for LCDR Skip Umstead, the Blue Angels "Boss."

United States Navy Blue Angels Flight Helmets

On 26 July 1973, while performing pre-arrival maneuvers in Lakehurst, New Jersey, "Boss" Umstead, flying the #1 jet, collided with Captain Mike Murphy, flying the #4 "Slot" position at about 1,000 feet above the ground. Both aircraft spun into the ground. Capt. Murphy and his crew chief, PO1C Ronald Thomas both ejected from their aircraft but did not survive. Sadly, LCDR Umstead did not eject prior to his aircraft impacting the ground and was killed in the accident. PO1C Harvey successfully ejected and was treated at a local hospital for minor injuries. He was wearing the helmet on this page.

This beautiful example is enhanced with silver metal flake, a trademark of Protection, Incorporated, a subsidiary of GENTEX and producer of custom flight helmets for the Blue Angels. The Blue Angels full color "crest" decal is centered on the visor housing. In more contemporary versions, the same dark blue graphics used by the pilots are used on the crew chief helmets. The result is a subdued appearance, with the only exception being their name or position where the "call signs" are located on the pilot helmets. The "crests" were also repositioned on the rear of the flight helmet, identical to the pilot's helmets.

This tragic accident ended the Blue Angels' 1973 show season and the Team was subjected to much scrutiny from many who thought they should be disbanded. Secretary of the Navy, John Warner, ordered the Team to stand down to undergo a review by a panel of six senior flag officers, who believed the Team were a "prime recruiting asset" unanimously recommended the Blue Angels continue their mission. Chief of Naval Operations, Admiral Elmo Zumwalt, accepted the recommendation but also requested a reorganization of the Team, including a replacement for the F-4J "Phantom."

U. S. Navy photo

The Navy's frontline fighter — the F-14 "Tomcat" — was given consideration, but determined to be far too costly to maintain and operate, and had not yet been formerly introduced into service. The Vought A-7 "Corsair II" was given a hard look, but couldn't be justified, given their mission in support of the U.S. involvement in Southeast Asia. At the end of the day, the U. S. Navy selected the McDonnell Douglas A-4F "Skyhawk" which entered service with the Blue Angels in 1974. This aircraft proved to be a good choice given the abundance of aircraft in inventory, was highly maneuverable, easy to maintain (in contrast to the F-4J), and economical to operate during the fuel crisis of the early 1970s.

In the U. S. Navy photo (above, courtesy of John Slezak), it appears that those lucky enough to receive a VIP / incentive ride with the Blue Angels during the F-4 era also wore a blue APH-6 dual visor helmet like those issued to the crew chiefs.

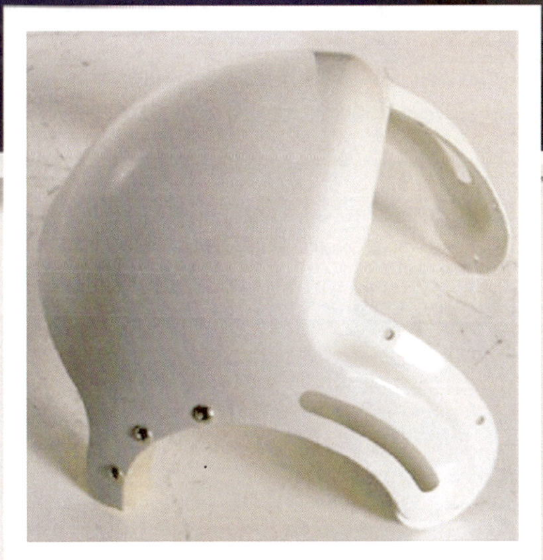

The dual visor kit assemblies were made by two companies, GENTEX and Sierra Engineering, each offering a unique approach to a common goal: Providing the crew member with a choice between a dark (neutral) visor for day operations, or a clear visor for night operations. GENTEX's version is what is known in collectors' circles as the "3-piece" visor kit, referring to the visor housing and the two "rams horn" housings for the visor actuating arms and lock assemblies (eBay photo - top). In contrast, Sierra Engineering offered a one piece approach moulded from a single piece of plastic and sized for medium and large shells (inset photo from flighthelmet.com).

The difference between the contemporary flight helmets issued to the crew chiefs and those worn by the pilots is simply the color of the paint. The pilots' helmets are painted yellow with dark blue graphics (Blue Angel crest on the rear and Blue Angel scroll on the visor housing), and the crew chiefs' helmets are painted blue — like the VIP helmets — with blue graphics.

In the example below, courtesy of John Slezak, this crew chief flight helmet is identified as "7 Crew Chief." This excellent example was made in 1996 but, bears no assigned name.

A Blue Angels Crew Chief meticulously cleans a pilot's flight helmet prior to a sortie. Regardless of whether this flight is a scheduled demonstration, practice show, VIP flight or maintenance hop, no detail is too small to insure the safety of the crew, passengers and aircraft. These men and women are hands down the "best of the best" the U. S. Navy and U.S.M.C. have to offer.

No detail too small.

Ground crew personnel are responsible for conducting comprehensive pre-flight inspections of each aircraft prior to all flights. These inspections include the pilots' flight gear. Any nicks or scratches on the gold mirrored lenses — however minor — would cause a distracting "starburst" effect in the sunlight and become a safety of flight issue, requiring immediate replacement of the lens.

U.S.M.C. Blue Angels?

When we think about the U. S. Navy Blue Angels, we often forget that the United States Marine Corps (USMC) has been an important component of the U. S. Navy since 1834. The USMC not only flies, crews and maintains the Blue Angels' C-130J support aircraft affectionately nicknamed "Fat Albert" but, they also fill one of the numbered demonstration pilot positions, as well as provide critical ground crew members in various technical capacities.

I want to be clear that the "Ground Crew" is a far larger and much broader entity than what I have captured in this chapter. Comprised of 15 specialty work centers staffed by approximately 120 sailors and marines, the effort is, in a word, massive.

It would be difficult — if not impossible — to capture the countless contributions of each work center. I venture to say that an entire book could be dedicated to sharing the remarkable work performed by these men and women day in and day out. Their tireless and selfless work ethic are what guarantees that the jets are ready and safe to fly, so you and I can enjoy witnessing their awe inspiring aerial ballet. They work 24 hours per day, 7 days per week to successfully launch an air show that lasts 30 to 45 minutes — the epitome of dedication. I have a tremendous amount of respect for, and salute these fine people:

Crew Coordinators
Career Counselor
Administration
Airframes
Aviation Medicine
Crew Chiefs
Events Coordination
Fast Albert Maintenance
Life Support
Maintenance Control
Supply
Paint Shop
Power Plants
Public Affairs
Quality Assurance
Video

The photos on this page were taken by the author inside the Blue Angels Life Support Shop, aboard NAS Pensacola, July 2021.

There is only one flying crew chief on the Team, and is assigned to the #7 jet. This is the helmet of the primary crew chief for the #7 jet flown by LT Julius "Whiskers" Bratton. The crew chief — referred to as the "7-Geek Alpha" is SSgt. Nathan "Rocky" Lyons.

These helmets are following the paint and graphics liveries of the VIP / PA flight helmets — yellow graphics on blue paint.

A beautiful example of a very rare Blue Angels flight helmet.

17
Blue Angels Life Support

To be honest, I had some doubts this chapter would get written but, in the end, persistence paid off. To put it lightly, securing an on-site visit to the Blue Angels squadron Life Support Shop (this, my third visit) was a tough nut to crack. Further complicating the matter were two unforeseen and powerful forces compliments of calendar year 2020: COVID-19 and Hurricane Sally. Once the pandemic and Mother Nature settled down, site visit approvals began to fall into place.

As I mentioned in the Introduction and Chapter 16 "The Ground Crew", I had the honor and distinct privilege of visiting the Blue Angels Life Support Shop on two previous occasions, 1997 and 2018. Of course, this visit would require laser focus in order to get the information and photos I needed all the while minimizing my disruptive impact on the Blue Angels' OPTEMPO[1].

In June of 2021, I received a phone call from from an unknown number. Thinking it was a telemarketer or SPAM, I declined the call. I immediately received a second call from the same number, so I reluctantly answered, and happy I did. It was LT Chelsea Dietlin, Blue Angels Public Affairs to discuss protocol and potential dates for a squadron visit. I was stoked! LT Dietlin cut to the chase and asked if I had reached out to the Navy Office of Information - East (NAVINFOEAST)[2] for support. I did but, had not received a reply so LT Dietlin helped me to pave the way, as I would be required to complete a formal request for author support.

Once the formalities were satisfied, we focused on a visit date.

---

[1] Operations Tempo: The rate at which units of the armed forces are involved in military activities, including contingency operations, exercises, and training deployments. It is time away from homeport or permanent duty station at the unit level.

[2] The Navy Office of Information - East coordinates outreach events as well as provides support and guidance to authors throughout its U. S. Navy Author Support Program.

According to LT Dietlin's official Blue Angels biography, she hails from Rochester Hills, Michigan, where she graduated from Rochester Adams High School in 2008 and earned her Bachelor of Arts degree in science, technology and environmental policy from Michigan State University in 2012. In 2016, Lt. Dietlin earned her commission as an Ensign in the U. S. Navy through the Officer Candidate School in Newport, Rhode Island.

LT Chelsea Dietlin, Blue Angels Public Affairs Officer. (U. S. Navy Blue Angels photo)

LT Dietlin joined the Blue Angels in 2019. Her decorations include one Navy and Marine Corps Commendation Medal, Navy and Marine Corps Achievement Medal and various personal and unit awards.

The day had finally arrived and I could barely contain my excitement. I planned to drive to Pensacola, Florida from my home in Lafayette, Louisiana — about a 5-hour trip — the first week of July 2021. My initial purpose was to attend the Blue Angels Pensacola Beach Air Show — also referred to as Beach Week — scheduled for the weekend following the 4th of July. If you haven't attended, I highly recommend it!

While in Pensacola, I also scheduled an in-person interview with Major Jeff Mullins, USMC, who flew with the Team in 2018 and 2019.[3]

[3] See Chapter 13 "The Yellow Helmet"

"Lady Luck" looked favorably upon me. LT Dietlin was trying to accommodate my squadron visit while I was in town! This was a huge win, and I was particularly excited because in addition to this being "Beach Week," it is also known as "Rush Week,"[4] where the Blue Angels welcome the top U. S. Navy and USMC finalists who are being considered for open positions on the Team. Those candidates "rushing" the Team undergo intense scrutiny through a series of formal interviews and social commitments carefully orchestrated to identify the right person for the right position — both on and off the job. The days and hours are long. The competition is intense. In the end, the Blue Angels officers — a total of 16 — must unanimously vote on who will be invited into the fold. My hope was to be lucky enough to breathe some of this rarified air!

Once again, Mother Nature showed us her demented sense of humor. The morning I departed for Pensacola, Tropical Storm "Elsa" — with sustained winds of 65 mph — was snaking her way northward in the Gulf of Mexico along the Western Florida coast. The good news was she was forecast to take an easterly track and head northeast into the Atlantic Ocean with little or no impact to Pensacola. With fingers crossed, I departed for Florida.

But…before I could ever leave my driveway, I received an email from LT Dietlin asking if I could make my squadron visit the week *after* "Beach Week?" My first choice would have been to extend my stay, but I had some personal commitments that required my return to Louisiana; however, given the circumstances and challenges of this visit, I was willing and able to make a quick turn. And, I did just that.

The following week, I departed early on a Monday morning to position and prepare myself for a Tuesday squadron visit. LT Dietlin advised me that MC2 Cody Hendrix[5] would facilitate my visit. I was genuinely excited — and humbled — to be granted this unique "inside" access and planned to make the best of my brief time there.

This is going to be epic!

---

[4] See Chapter 12 "Newbies"

[5] Mass Communication Specialist.

About the Author

I have spent a great deal of my adult life fascinated by, and studying the science and art that is known as aviation life support. As a collector and historian, I have amassed a respectable collection of military flight gear, including, but not limited to flight helmets, oxygen masks, flight suits, ejection seats, instrument panels and more.

Over time – through technical manuals and informal training from PRs, AFEs and ALSE Technicians – I learned about the maintenance and repair of life support equipment, which came in handy conserving my growing collection. I fancy myself something of an amateur life support "hobbyist," if you will.

As my curiosity and thirst for knowledge grew, I began dabbling in flight helmet restoration, and it quickly became my biggest passion – a passion that has grown into a small niche business providing specialty (non-airworthy) flight helmet restoration services for collectors across the globe.

The early morning road trip was uneventful and I checked in with MC2 Hendrix upon my arrival. We exchanged phone numbers and agreed to meet at 9:00 a.m. on Tuesday morning at a pre-arranged location just outside of the NAS Pensacola main gate.

On Tuesday morning, at precisely 0900 hours, MC2 Cody Hendrix arrived at the appointed place and escorted me onto NAS Pensacola. It has been many years since I accessed the north gate and I had almost forgotten how beautiful the base was from this vantage point.

Upon arrival, I was escorted directly to and handed off to the Life Support shop where I would be taken good care of by PR1 Darby Versluis[6], AME1 Mario Acosta and PR2 C. J. Woods. I would later be introduced to the shop supervisor AME1 Jack Reynolds, a fellow Cajun from South Louisiana! AME1 Reynolds is in his 4th year with the Blue Angels and will remain on for an unprecedented 5th year for the 2022 show season prior to retiring from Navy service!

As previously stated, I was privileged to visit the Life Support shop on two prior occasions, 1997 and 2018 but, these

[6] Promoted from PR2 to PR1 the day of my visit.

visits were brief. This visit afforded me an opportunity to take in the space and all that it had to offer. The shop was modest by any definition. This reaffirmed to me that the quality of work produced here is a result of talented and dedicated sailors, not from some luxurious work space. The shop was functional and probably not unlike the space available on the boat.

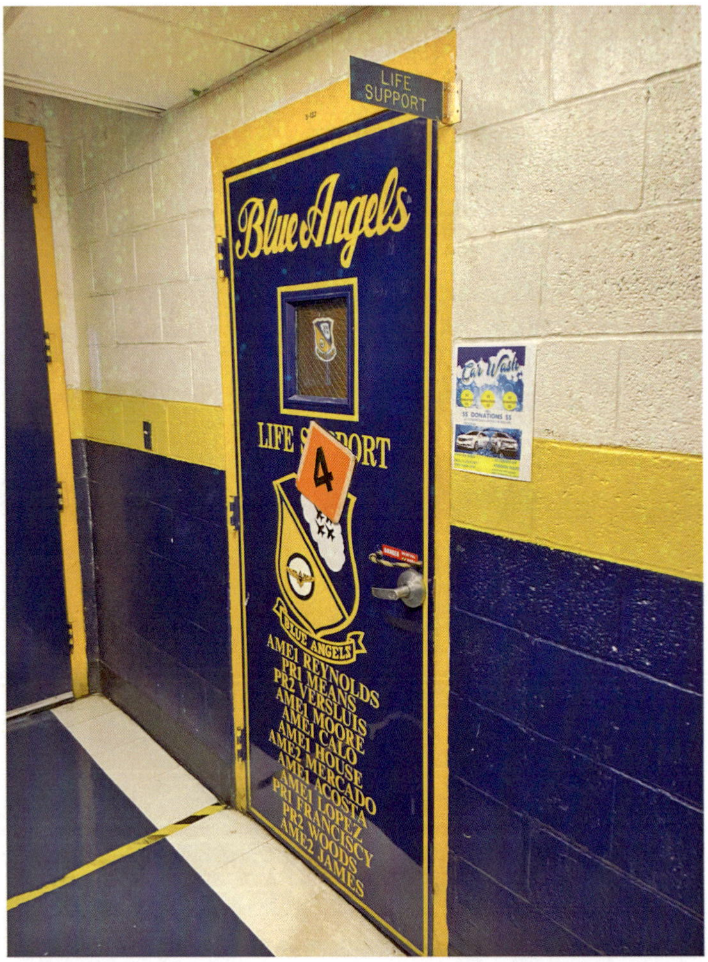

Blue Angels Life Support Shop, NAS Pensacola. (Author's photo)

Once inside this "inner sanctum" my goal was to make the best use of my time and be the least disruptive as possible but, because these sailors were working there were many work-imposed "interruptions." As is the "norm" for this place the OPTEMPO was high and since it was Tuesday, the Team had a scheduled rehearsal. The pace was exhausting!

Speaking of rehearsals, I was invited to watch the Team practice just outside of the Blue Angels' hangar...but, I digress!

I had many questions, most of which were technical and considered by many to be "in the weeds," but, that's where many of us live and thrive. I inquired about their interface with Gibson & Barnes, the single-source provider for their custom flight helmets. I learned that orders are placed as new pilots are assimilated onto the Team based on standard shell sizes (e.g. medium, large and extra large). The helmets are delivered to the Blue Angels complete, less communications and the iconic gold visor. The Life Support shop

replaces the standard-issue dark visor with a gold mirrored visor sourced from Oakley, best known for their high-end sunglasses. Custom comms are also installed. Due to the high noise generated by the new F/A-18 "Super Hornets," the Team has experimented with, and will be introducing a new comm setup that includes active noise reduction. The pilots are stoked!

The photo above and to the right are PR1 Darby Versluis and PR2 C. J. Woods performing comms checks on a couple of Blue Angels flight helmets. The helmets are checked when they arrive from Gibson & Barnes, and on a routine basis to insure the pilots have the best communications under very demanding conditions.

I was shocked to learn that in the Life Support Shop — which is not much bigger than what you see in the two photos on the previous page — the AME's also work on ejection seats! This gives a new meaning to tight quarters but, like everything the Blue Angels do, they make it happen…every day!

Technically, the Life Support rate is known as Aircrew Survival Equipmentmen. The following is from to the U. S. Navy Web site[7]:

As an Aircrew Survival Equipmentman (PR), you're responsible for maintaining the safety equipment that could save a Naval Aviator's life in the event of an emergency. Your job responsibilities as a PR may include:

- **Inspecting, maintaining, and repairing survival equipment, flight gear and protective clothing.**
- **Inspecting and testing oxygen regulators, converters and related equipment.**
- **Rigging, packing and repairing parachutes.**
- **Equipping and packing life rafts.**
- **Operating and maintaining sewing machines and performing fabric work necessary for survival equipment.**
- **Volunteering for aircrew duty.**

Coincidentally, the Aircrew Survival Equipmentman course — 12 weeks — is held at NAS Pensacola. The Navy Web site suggests that PR applicants should possess good writing and speaking skills as well as an aptitude for working with tools, equipment and machines. Applicants should also be able to perform meticulous detail work, maintain accurate records and work well as part of a team. Important personal attributes as a PR include resourcefulness, curiosity, and the ability to follow through on tasks. Attention

[7] https://www.navy.com/careers/aviation-survival-equipmentman#

to detail and good memory are vital, as your work will directly impact the lives of other people. You should also be physically fit and have normal use of your hands.

The PR creed is:

"The last ones to let you down."

Powerful words to live by…day in and day out. In this business, failure is not an option.

(Author's Photos)

My sincere gratitude goes out to this exceptional team of sailors who gave generously of their valuable time to welcome me into their shop and educate me on their critical mission. There were many more who supported my visit — especially the crew chiefs — to whom I was not formally introduced. Thanks to all of the unsung heroes — you are appreciated!

AME1(AW)

Jack Reynolds

Moss Bluff, Louisiana

PR1(AW/SW)

Darby Versluis

Neosho, Missouri

AME1(AW)

A. J. Acosta

San Angelo, Texas

PR2(AW/SW)

C. J. Woods

Dallas, Texas

18
Pensacola Naval Air Station

United States Navy Blue Angels Flight Helmets

Often referred to as the "Cradle of Naval Aviation," Pensacola Naval Air Station (NAS) is the primary training base for all U. S. Navy, U. S. Marine Corps and Coast Guard officers who wish to become Naval Aviators and Naval Flight Officers. It is also the advanced training base for most Naval Flight Officers as well as home of the United States Navy Flight Demonstration Squadron the Blue Angels.

Pensacola NAS is a magical place that lures me back time and time again, despite having visited more than a dozen times. Located along the spectacular "Emerald Coast," Pensacola NAS has been a favorite duty station for naval aviators for decades. As an added bonus, the National Museum of Naval Aviation, a world-class facility by any definition, is also located on base.

The research for this book was more than enough reason to visit this beautiful lady one more time!

I made a return visit to Pensacola NAS the Fall of 2019. This trip afforded me the opportunity to reminisce on previous memorable visits, capture some photos I missed and, of course visit, the National Museum of Naval Aviation!

Unfortunately, this trip was bittersweet. I was reminded of many good memories from previous visits, as I sought out new exhibits. Two of the Blue Angel flight helmets that I wanted to photograph had been removed from display and placed in storage, but as it turned out, this was the least of my concerns. In December of 2019, tragedy struck Pensacola NAS Three people were killed and eight injured when a foreign student opened fire from a handgun into classrooms of unsuspecting instructors and students. Following this incident, the base was closed to all personnel without a valid military ID — that included the National Museum of Naval Aviation.

It is said that time heals all wounds and I am confident that in time Pensacola NAS and the National Museum of Naval Aviation will once again welcome civilian visitors. I am especially hopeful this temporary restriction will be lifted in time for the annual Blue Angels Homecoming air show in October.

When most people think about Pensacola, the beaches typically come to mind. They are without a doubt very beautiful and memorable. When I think about Pensacola, the Blue Angels immediately come to mind...every time, without fail. For me and countless others, the Blue Angels are magical and epitomize the spirit of aviation.

I have enjoyed watching the Blue Angels perform for close to 50 years and continue to be filled with excitement in anticipation of each and every show I attend.

The Blue Angels Homecoming air show in 2020 was scheduled for mid-October, about 3 weeks earlier than their normal schedule of early November. This is to allow the Team a few extra weeks of preparation at NAF El Centro as they transition to the F/A-18 "Super Hornets" for the 2021 season that coincides with their 75th anniversary. This will be the last show performed in the F/A-18 "Hornets" that served them well since 1986.

The 2021 show season also began later than in previous years. Instead of March, the first Blue Angel show was in April. The shortened 2020 season coupled with the delayed start of the 2021 season will provide the Team with a few extra weeks to hone their skills in their new steed.

These are exciting times, and I was looking forward to witnessing the last Blue Angels show in F/A-18 "Hornets" as well as the first show in the F/A-18 "Super Hornets." As a self-admitted air show "junkie" this is without a doubt one of the most exciting times I can remember in my (aviation) lifetime.

Although I kept my fingers crossed that the weather would be favorable for both of these historic air shows because, as we know all too well, Mother Nature can be a cruel mistress. But as we drew closer to that long awaited October Homecoming airshow, the Corona Virus / COVID-19 pandemic of 2020 struck Florida particularly hard, and the Blue Angels did the only thing they could do, cancel the entire 2020 airshow season. As difficult as this news was to accept, I had to continually remind myself that the decisions surrounding the cancellations were designed to insure that we would have future opportunities to be entertained. I get it, but it was a hard pill to swallow.

This dark cloud did come with a silver lining…of sorts. On 27 July 2020, CAPT Eric Doyle, Director of the Blue Angels Super Hornet Transition Team, and former Blue Angels "Boss" delivered the first Blue Angels F/A-18 "Super Hornet" to Naval Air Station Pensacola. As a tribute the to first Blue Angels "Boss" the aircraft exhibited the name "Capt. Roy 'Butch' Voris" on the canopy frame. BRAVO!

The photos on the next page are of the first F/A-18 "Super Hornet" delivered to NAS Pensacola by CAPT Eric Doyle to CDR Brian Kesselring, the current Blue Angels Boss who will be leading the 2021 Team in their new and improved aircraft. Here's hoping for a long and exciting 2021 airshow season!

U. S. Navy Blue Angels photos.

19
GENTEX

This book would not be complete without GENTEX. General Textile Mills — now doing business as GENTEX — is the largest manufacturing plant specializing in high tech protective headgear. Located in Simpson, Pennsylvania, GENTEX's primary customers are U.S. and international militaries as well as law enforcement and civilian markets. Notable customers include the USAF Thunderbirds and, of course, the U. S. Navy Blue Angels. In fact, the H-1, H-2, H-3, H-4, DH-151, HGU-33/P (PRK-37/P shell) and the current HGU-55/P flight helmets were all produced by GENTEX in the same facility. A true testament of quality.

When the United States entered the second World War, General Textile Mills began manufacturing parachutes for the U.S. Military, and eventually became one of the largest manufacturers in the country. During that same period General Textile Mills produced composite-structure parachute boxes from fiberglass mat impregnated with a polyester resin that would protect packaged parachutes. Following the war, the government approached General Textile Mills to determine the feasibility of using the same composite-structure to manufacture rigid pilot helmets. Due to the advancements in aircraft technology, improved protective headgear was warranted to replace the cloth or leather options used prior to 1948.

General Textile Mills produced its first hard shell helmet using the parachute box technology that same year. The helmet, known as the H-1 was made for the U.S. Navy — and used by the Blue Angels — and the P-1 soon followed for the United States Air Force. These technologies were quickly incorporated into helicopter pilot helmets for both military and commercial applications as well as tank crew helmets.

In 1958, General Textile Mills shortened its name to GENTEX, and the company continued working to improve and adapt the helmet technology.

Significant advancements in the aircraft industry resulted in the need for oxygen-breathing equipment for life support. In 1969 GENTEX bought Protection, Inc., a helmet manufacturer operating as a division of Mine Safety Appliance (MSA) in Pomona, California. Ten years later, Protection Inc. began to acquire equipment in order to manufacture respiratory products. By 1981, the new branch officially transitioned into GENTEX Western Operations focusing on the manufacture of respiratory products. The California operation moved to Rancho Cucamonga in 1992.

In 1983, GENTEX sold the custom-fit helmet business, doing business as Protection, Inc. to Flight Suits, Ltd., doing business as Gibson & Barnes.

GENTEX is a name synonymous with military flight helmets and a GENTEX sales brochure sums it up: **"If you've got the right stuff, wear the right stuff.** When it comes to flying aviation pros know it pays to own and use the best equipment money can buy. In protective helmets, that means Gentex, the name synonymous with quality, innovation and service to pilots. Carefully designed, constructed and tested for the ultimate in comfort and safety, Gentex helmets help you fly your best no matter how challenging the conditions. Available in off-the-shelf or custom-fitted and painted models. Gentex helps you show your stuff."

While conducting research for my book, *The Red Helmet*,[1] I contacted Mark Jones, GENTEX Aircrew Helmets Product Manager, who arranged for an on-site tour of their manufacturing facility.

The GENTEX Plant is located in an unassuming building and nestled among an established neighborhood on Main Street in Simpson, PA. Visitor parking is across the street next door to a craftsman style home typical of the area of less than 1,300 residents.

The tour began with a brief oral history of The General Textile Mills - GENTEX - just inside of the sprawling 100+ year old, multi-story plant. The wooden floors were original to the building, well worn, and scarred from several generations of families who have worked here since before World War One. I learned from talking with several employees that 40 to 45 years of service is not uncommon. A true testament to a successful, well run, family business that recognizes the importance of a loyal workforce.

As we began what turned out to be a 2 1/2 hour tour, I was most surprised by the very limited use of automation and the lack of a "production line" that likely comes to mind when manufacturing anything in mass quantities. Instead, there were dozens of small, informal work stations scattered throughout the plant where hundreds of dedicated employees hand-build every single flight helmet. This is difficult to wrap your mind around considering GENTEX makes approximately 9,500 flight helmets annually. One at a time.

GENTEX produces complete, made-to-order flight helmets sold directly to the United States Navy and United States Air Force, just to name a few from an extensive customer list. Flight helmets are also sold to the law enforcement and civilian markets through their authorized dealer, Gibson & Barnes in El Cajon, California. In the case of the Blue Angels, these orders are very specific. Let's take a look.

---

[1] Elm Grove Publishing, ISBN 978-1-943492-76-3

United States Navy Blue Angels Flight Helmets

The photo on the top left is an HGU-55/P shell as it appears after being unfolded, trimmed, drilled and primed.

If this shell was destined for a U. S. Military contract customer, the shell would be painted and a host accessories (depending on the customer's requirements) would be added. Typical accessories may include a hand-sewn leather edge roll, a Thermo Plastic Liner, chin and nape straps, ear cups, earphones, and an external visor. The finished product would like similar to the photo in the center.

The helmets destined to become Blue Angels helmets take a decidedly different path.

The Blue Angels source their custom flight helmets from Gibson & Barnes. As new pilots assimilate into the Team, Blue Angels Life Support personnel order the requisite number and size helmets to meet their current needs. Gibson & Barnes places their order with GENTEX and receives shells only, with only a coat of primer and stripped of any accessories.

This is where the magic begins. Artists and painters at Gibson & Barnes paint the naked shells Blue Angels yellow. While in the paint booth, they also paint an EEK-4A/P visor housing and visor lock knob. Once the paint cures, Blue Angels graphics are applied: The Blue Angels "scroll" centered on the forward edge of the visor housing and the Blue Angels "crest" centered on the rear of the shell. The pilot's call sign is also applied centered on the back of the shell just above the edge roll.

The shell and visor housing is then covered with multiple applications of clear coat followed by wet sanding and polishing to a mirror finish. A hand-sewn black — not government-issued grey — edge roll is added as well as black, leather covered chin and nape straps. These unique HGU-55/P flight helmets are further "disguised" with the addition of the iconic, gold mirrored visor and chrome plated oxygen mask receivers. The transformation is nothing short of breathtaking!

GENTEX. A time-honored name you can trust. Since World War I, several generations of families have dedicated their lives to producing world-class protective equipment for military, law enforcement and civilian pilots across a global enterprise. A sign hanging proudly in the 100+ year old plant reads:

"GENTEX: Lives Depend on Us."

GENTEX
CORPORATION

20
Protection, Incorporated

This story would not be complete — or possible — without a company many have never heard of, Protection, Incorporated (PI) of Pomona, CA. The brain child of Dr. Charles Lombard, PI was the leading provider of custom flight helmets — most notably, to the Blue Angels[1] for many years. Not much is known about PI, but fortunately we can gain a little insight from the following statement of Company Qualifications[2] in an undated archival document:

"Protection, Inc., is a corporation organized and existing under the laws of the state of California. The company was originally organized in 1947 as a California corporation and has a continuous history since that time in the design, development, and production of aircrew and personal protective equipment. In the period from January 1, 1957 to September 30, 1960, Protection, Inc. was operated as a division of Mine Safety Appliance of Pittsburg, PA. As of October 1, 1960, its status was changed to a privately-owned California corporation and has continued as such to date."

In 1969, the custom-fit helmet business of Protection, Inc. was purchased by GENTEX and subsequently sold in 1983 to Flight Suits, Ltd., now doing business as Gibson & Barnes.

In the summer of 2019, Mark Jones, Aircrew Helmets Product Manager at GENTEX, introduced me to a former PI employee, Lucien Van Oosten, who was working as a Senior Engineering Specialist for GENTEX in Southern California. While employed by PI, Mr. Van Oosten was part of the flight helmet design engineering team, but credits Dr. Charles Lombard and Robert Zimmerman as the "pioneers" in the custom fit flight helmet business for USAF test pilots, Lockheed, Douglas Aircraft and many more.

---

[1] Protection, Incorporated also made custom flight helmets for the USAF Thunderbirds.

[2] From the personal archives of Mike Baetslé, Belgium

In 2020, I became acquainted with Steve Norris, a fellow researcher, flight helmet collector and historian who knew Robert Zimmerman.[3] Employed by Protection, Inc. and GENTEX from 1953 to 1998, Mr. Zimmerman dedicated his 45-year career to designing, developing and manufacturing custom fit flight helmets for the U. S. Navy, U. S. Air Force, and NASA.

When Mr. Zimmerman began working for PI, he was asked by management to become "...acquainted with modeling clay, wood modeling...machining of metal parts, working with fiberglass, polyester and epoxy resins, forming acrylic visors, creating forms for forming acrylic, getting educated on hardware used in helmets, etc."[4] This predates the use of computers and underscores the challenges associated with creating and building high-tech products using the power of the brain, and the skills of the human hands — a dying art today.

PI is probably best known for their model T-3 flight helmet — the TOPTEX — sometimes referred to as the Lombard helmet, in honor of the principal designer, Dr. Charles Lombard. Mr. Zimmerman estimates that between 1949 (TOPTEX

[3] Robert Zimmerman, deceased

[4] From a letter written by Robert Zimmerman to Steve Norris.

patent application) and 1982 (end of production), approximately 15,000 TOPTEX helmets were produced for the U. S. Navy, U. S. Air Force, U. S. Marine Corps and some foreign governments.[5] Although the USAF Thunderbirds chose the TOPTEX helmet for their use primarily in the 1950s, the Blue Angels opted for standard issue helmets of the period.

From these humble beginnings, Mr. Zimmerman went on to custom fit the original seven Mercury astronauts with TOPTEX helmets for use while flying their NASA T-38s, worked on designing their space helmets, as well as worked on the contract for the Space Shuttle helmets, which was a modified version of the HGU-20/P.[6]

Protection, Inc. APH-6 flight helmet liner tag from Blue Angels pilot Dick Schram, 1969. *(Courtesy of John Slezak)*

Adding to this impressive resume, PI also provided custom flight helmets to the U. S. Air Force Thunderbirds from 1955 to 1982, as well as the U. S. Navy Blue Angels from 1967 through 1982.[7] During this 15 year period, we witnessed several flight helmets used by the Blue Angels, including the APH-6, DH-151, and the HGU-33/P, the latter of which is somewhat of a misnomer.

According to James Wegge, the Director of Gibson & Barnes, the industry and collecting circles have often referred to the Blue Angels flight helmets built on a PRK-37/P as an HGU-33/P, but the Blue Angels version had different components than a U. S. Navy-designated HGU-33/P, therefore, was technically not an HGU-33/P. The Blue Angels switched over to the contemporary HGU-55/P[8] in the mid-1990s.

---

[5] From a letter written by Robert Zimmerman to Steve Norris.

[6] Ibid

[7] Ibid

[8] Fitted with an EEK-4A/P visor housing kit and re-designated HGU-55/E by Gibson & Barnes.

That being said, I have been involved in my share of online debates — with like minded collectors across the globe — regarding the authenticity of a flight helmet, or the proper alpha-numeric designation of a bare shell vs. a complete helmet, or perhaps arguing the finer details on the differences between a U. S. Air Force flight helmet and a U. S. Navy flight helmet. The majority of these "information exchanges" have been civil and educational, as one should expect; however, some ended in stalemates, and, unfortunately, a few ended in bruised egos by one party or another.

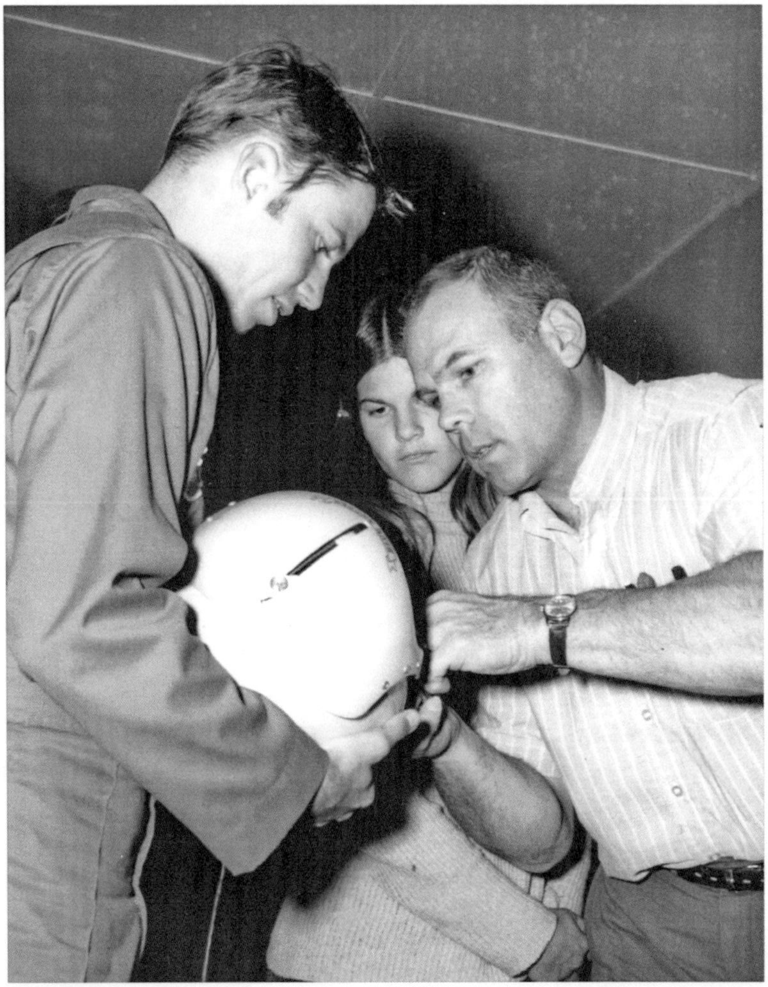

Robert Zimmerman, Protection, Inc. (right) with Blue Angel pilot Jim Maslowski, 1972. Unidentified female.
(Photo courtesy of Steve Norris)

As a researcher, historian and flight gear collector for the past 40 years, I am fascinated by knowledge and still consider myself a student of this strangely satisfying obsession.

I learned a couple of things while researching this book that could put to rest some age-old arguments about flight helmets...or perhaps fuel new discussions.

In December 2018, I was granted a tour of the GENTEX plant in Simpson, Pennsylvania, led by Mark Jones, Aircrew Helmets Product Manager. Mark was somewhat familiar with the fervor of the flight gear collecting community, and suggested that many would be surprised to learn that many different flight helmets are formed on the same heat and pressure molds. The shells are then trimmed and drilled according to the customer's requirements.

Robert Zimmerman echoed this in a letter he wrote to Steve Norris. He said, "In 1958, the Air Force contracted with Protection, Inc. to create its own version of the Navy APH-5/P and APH-6/P helmets. The two helmets were essentially the same in design." He continues, "The Air Force and the Navy used the same shell at this time, but each one had a different designation. I believe that as of the time I retired [1998], both services were still using the same mold to laminate the shell that formed the basic helmet."

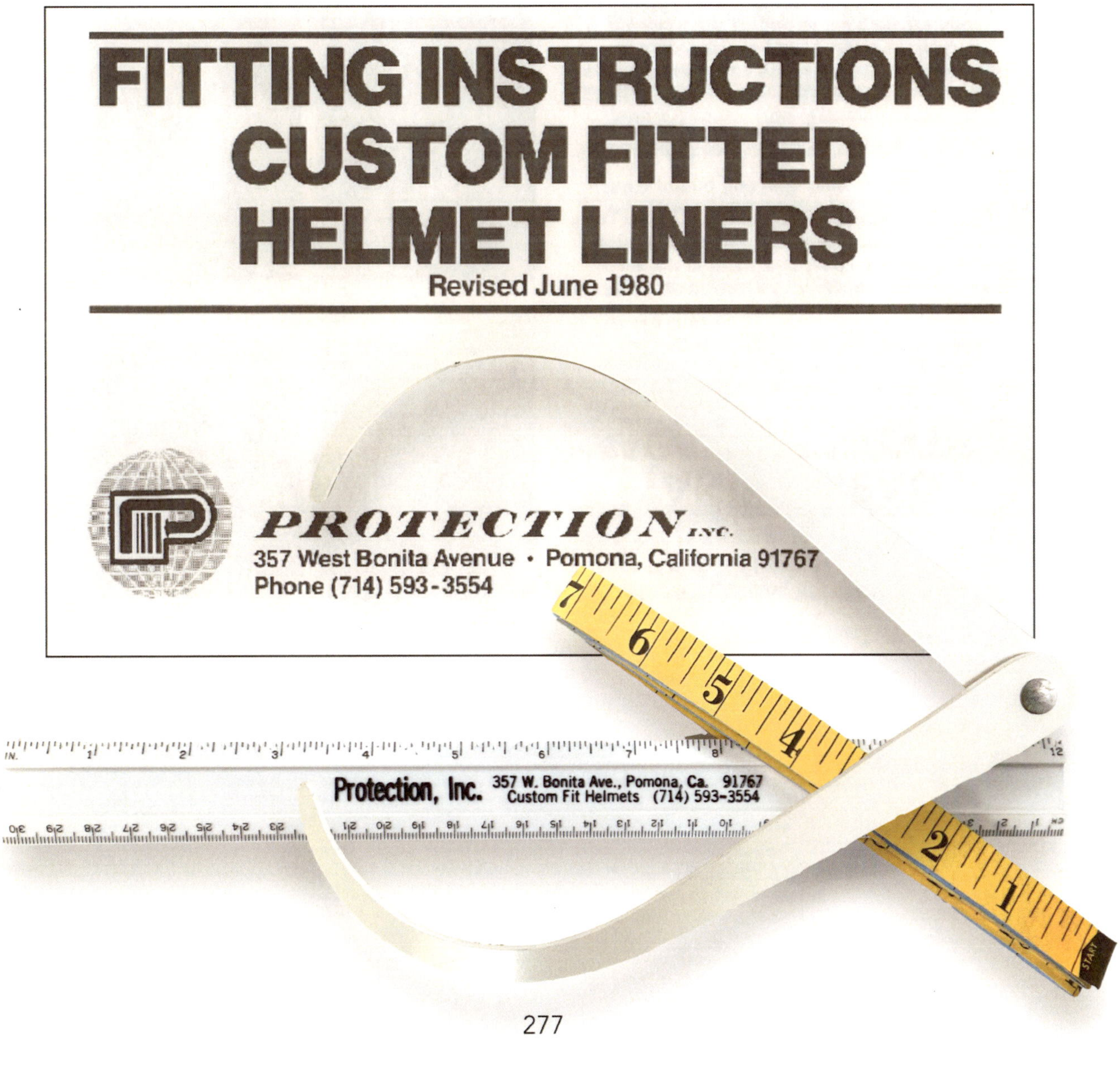

The Blue Angels' decision not to choose the Protection, Inc. TOPTEX — an arguably superior helmet over their GENTEX and Sierra Engineering government issue counterparts — may never be known. What we do know is Protection, Inc. secured the Blue Angels contract to custom paint and fit their flight helmets.

The photo on the previous page (author's private collection) shows the instructions and tools used to provide Protection, Inc. with the critical measurements of a pilot's head that are necessary to make a custom fitted helmet liner. Protection, Inc. offered customers two options: 1.) The tools and instructions could be sent directly to the pilot who would record the measurements on a custom order sheet and return it, or 2.) The pilot could physically visit the Protection, Inc. facility where he / she would be measured and fitted by qualified staff members. The first option required a dedicated person to perform the fitting measurements.

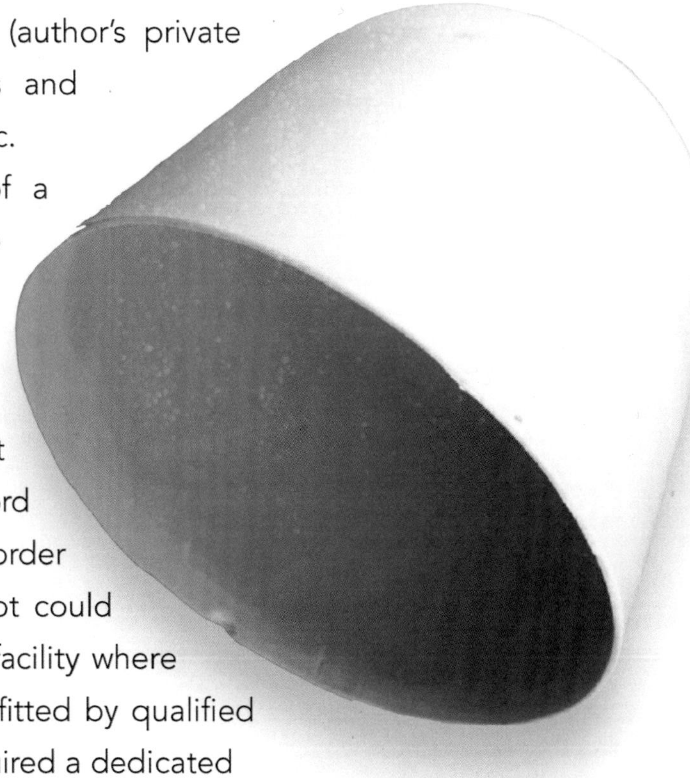

The photo above right (author's private collection) is an original, unused wax mold that was critical to the custom fit liner process. Due to the lack of records and fading memories, it is unclear which option the Blue Angels chose for their custom fitting process. Either way, the wax mold was placed in a container with water heated to 130 degrees F to soften the wax. Once pliable, the wax mold was placed on the pilot's head and allowed to cool. The mold creates a replica of the pilot's head, and is used by Protection, Inc, along with a host of physical head and face measurements to create a custom fit liner.

The detailed wax impression instructions and custom-fitted helmet order form follow (author's private collection).

MAKING A WAX IMPRESSION

1. Have water in container at about 130 degrees F. Hold wax mold in container by edge. Rotate slowly so all parts are heated uniformly. When wax mold is heated uniformly and is pliable, lift from water and drain out water. See Photo 1. Cracking of Wax Mold. If the wax mold cracks, it could be for any one of the following reasons:
   A. Wax was too cold.
   B. Wax mold was worked too fast onto the head.
   C. Wax mold was not removed from the head soon enough. As soon as the mold is marked, it should be removed from the head.
2. Place wax mold on individual's head. Work wax mold down with both hands working slowly to avoid splitting. Lift wax around edges next to the ears as it is worked down so ear lobes do not become folded. See Photo 2.
3. Continue working wax mold down. Align wax mold so it is even on both sides and in the front and back. Continue working wax mold down so that it contacts the head. See Photo 3.
4. Work wax with both hands until it fits all over the area above the hatband line. Make sure shell contacts head as closely as possible around ears. See Photo 4.
5. Along centerline of nose, scribe a vertical midline on the wax using a ruler and pencil. CHECK THIS MIDLINE FOR ACCURACY. See Photo 5.
6. Scribe a horizontal line on the wax at the eyebrow for reference. Scribe a second horizontal line one inch above the eyebrow line. Mark the name of the individual on the wax. See Photo 6.

   NOTE: Before removing wax, place a dot on rear center of head one inch above edge. Use caliper to measure from this dot to center of reference cross formed on forehead. This will be a check for stretching during mold removal. Don't change setting until check is made.
7. Place both hands at front and back edge of wax. Loosen gradually and lift off. Place upright on table and check front to back

2

3

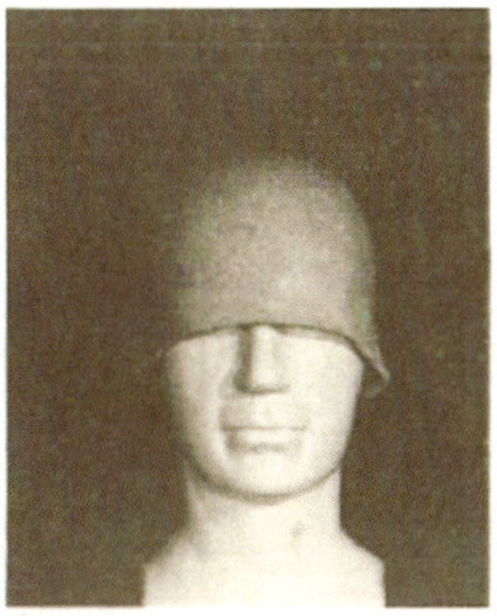
4

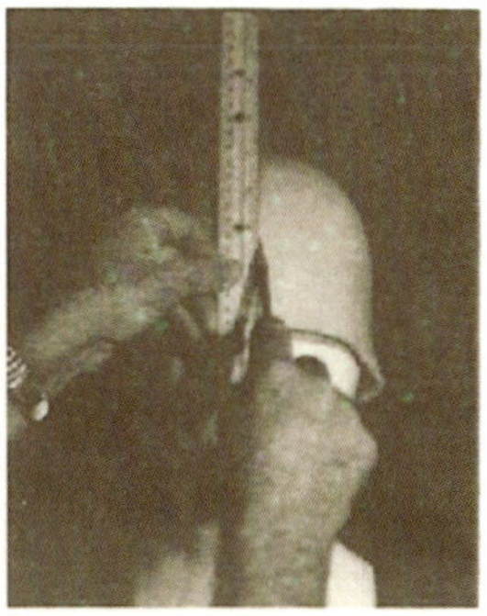
5

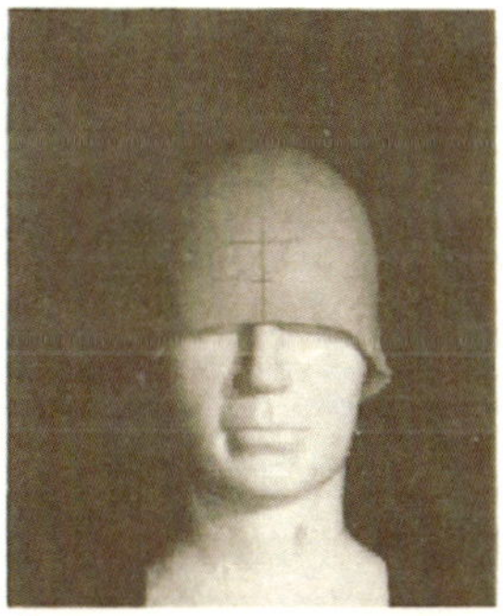
6

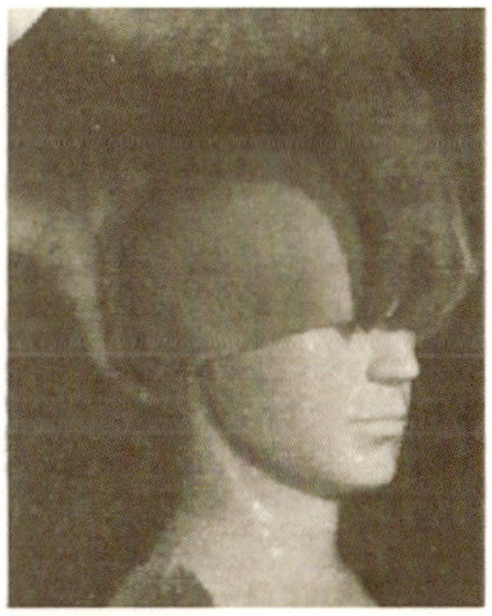
7

8

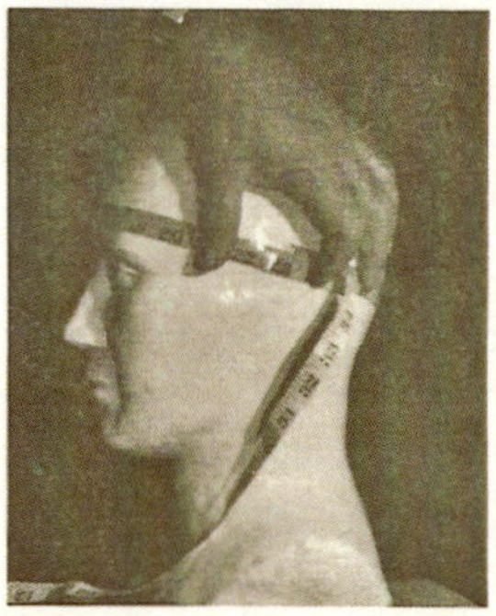
9

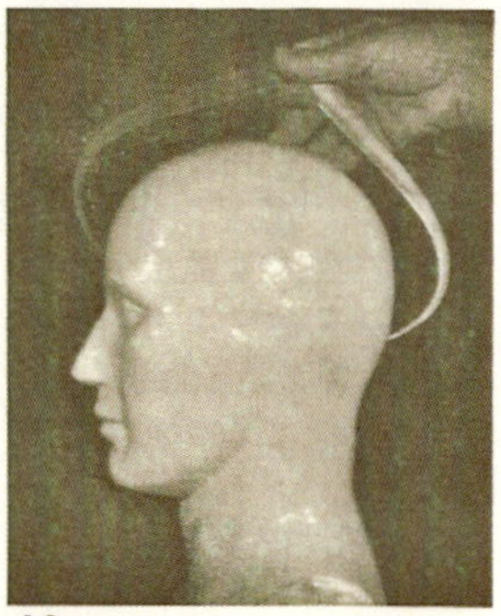
10

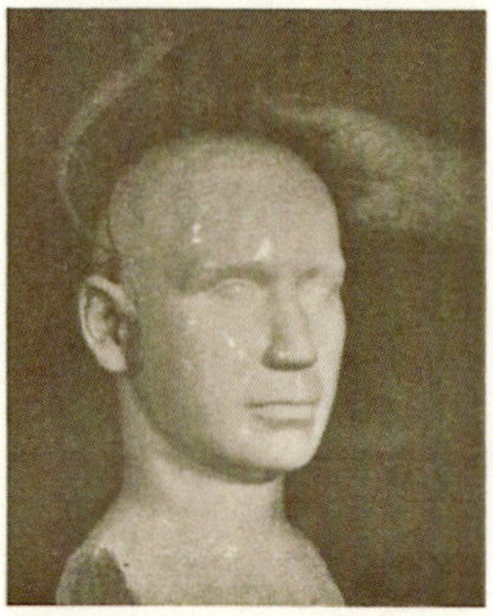
11

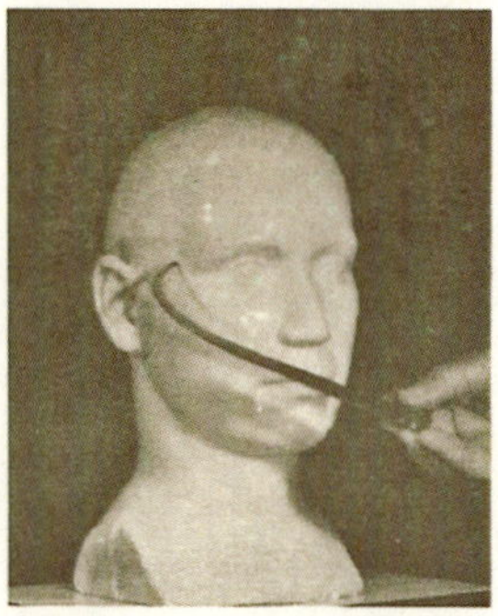
12

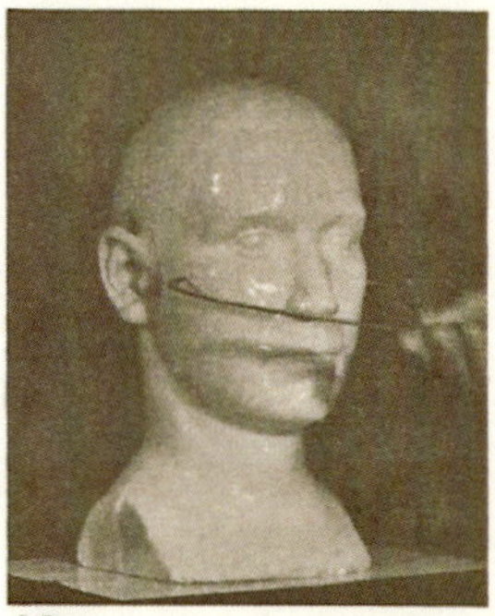
13

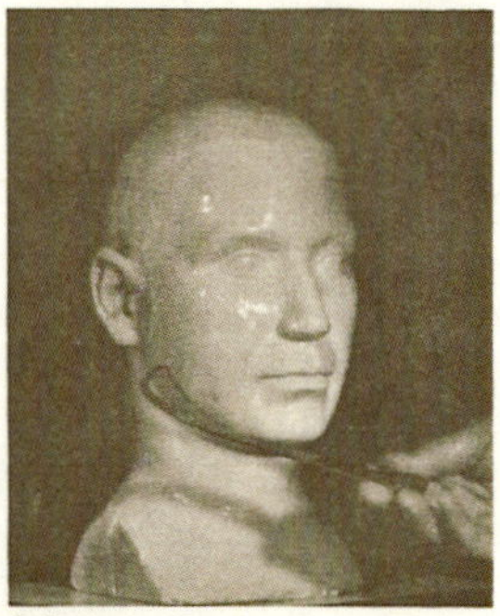
14

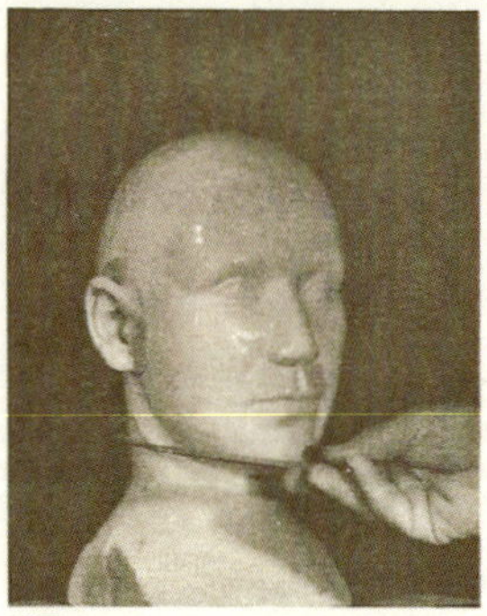
15

dimension with caliper. If it differs, carefully change to agree with setting as in paragraph 6. See Photo 7.

TAKING HEAD MEASUREMENTS

1. Record individual's name, address and other data on the fitting sheet. Also record the following dimensions on the fitting sheet where shown.
2. Measure distance from ear canal to ear canal over top of head with tape (all measurements to nearest 1/10″ or 1/16″). See Photo 8.
3. Measure the circumference of the head at the hatband line with tape. See Photo 9.
4. Measure head length at hatband line with calipers. See Photo 10.

   NOTE: On all calipered measurements, do not indent the surface of the skin; after the measurement is taken, use a ruler to measure the spread of the calipers and record this measurement on the fitting sheet.
5. Measure width of head over ears with calipers. See Photo 11.
6. Measure maximum width of head at temple with calipers. See Photo 12.
7. Measure maximum width over cheek bones with calipers. See Photo 13.
8. Measure maximum width over jaw bones with calipers. See Photo 14.
9. Using calipers, measure head width over rear angle of jaw bones.
10. With calipers, measure width of neck at midjaw point. Tip head up to take measurement.
11. Measure maximum width of neck – 1″ below ears, using calipers. See Photo 15.
12. Measure length of head from chin to top of head, with calipers.
13. Record if glasses are worn when wearing helmet and fill in all remaining data on the fitting sheet.

PROTECTION, INC.

357 WEST BONITA AVENUE • POMONA, CALIFORNIA 91767 • (714) 593-3554

CUSTOM-FITTED HELMET FITTING SHEET

NAME ______________________ SOC. SECURITY NUMBER ______________

ORGANIZATION ____

SHIP TO: ______________________ BILL TO: ______________________

ITEM DESCRIPTION: HGU-2A/P ☐ SPH-3 ☐ REWORK

HGU-33/P ☐ SPH-3B ☐ OTHER ______________

TOPTEX ☐ SPH-4 ☐

SIZE (To be determined by Protection, Inc.) MED LGE EX LGE MED LNG

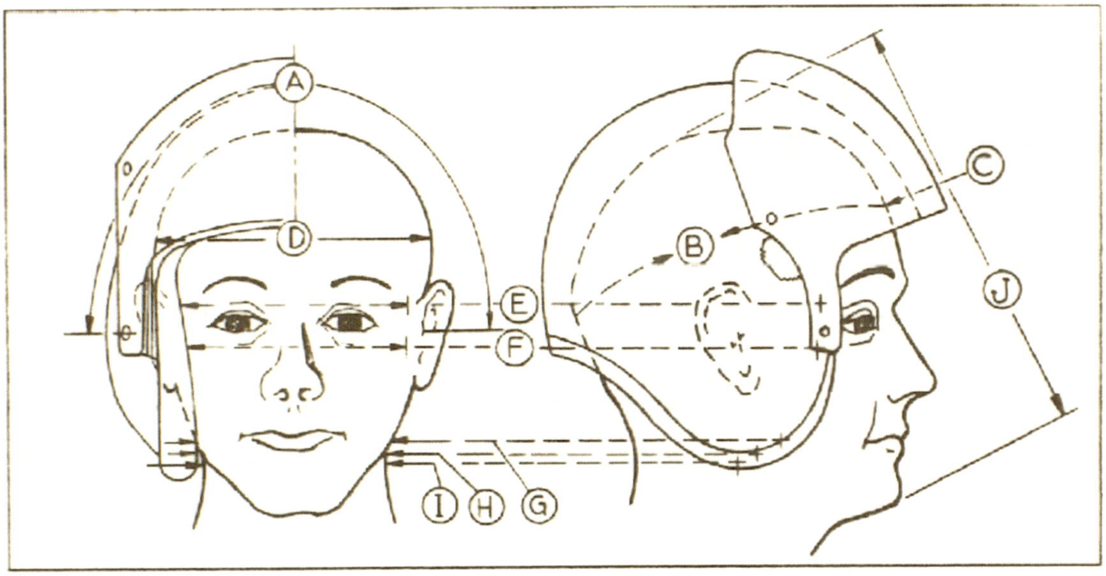

(A) Distance from ear canal to ear canal over top of head ______________

(B) Circumference at hatband line ______________

(C) Head length at hatband line ______________

(D) Head width over ears ______________

(E) Head width at temple ______________

(F) Head width over cheek bones ______________

(G) Head width over rear angle of jaw bones ______________

(H) Head width at midjaw and neck ______________

(I) Neck width 1 inch below ears ______________

(J) Head length chin to top of head ______________

Additional Requirements: Napestrap O_2 Mask Type________

O_2 Receivers, White Chrome Painted Color________

Visor, Single Dual Clear Dark Gold

Helmet Bag, Leather Nylon Large Small

Helmet Bag, Color ________ Extra Skullcaps, Qty____ Color ________

O_2 Bayonets, "T" Style Offset Style Navy AF

Communication Requirements: Earphones Type________

Cordset Type________ Microphone Type________

Adapter and Other ________

HELMET DESIGN: Shell Color________

Visor Color________

Liner Color ________ Edgeroll Color ________

Knob Color________

Will helmet be worn with glasses? Yes No If yes, indicate bow width ________

Name to appear on helmet ________ Front Back

Type of lettering ________

All Caps Upper and Lower Case Other________

DESCRIPTION OF DESIGN WORK (ATTACH SKETCHES) ________

REMARKS ________

Date of Fitting ________ Fitting Taken By ________

Terms of Payment: Credit Card Number ________

VISA M. Charge Expiration Date ________

Cash Check Amount Paid ________

CUSTOMER SIGNATURE ________

The vintage water slide decals on this page have been popular for many years but, not always used on the Blue Angels' flight helmets.

The top left decal is from the F-11 era and is courtesy of the John Slezak collection. It is believed that the Blue Angel "banners" below the crest were cut out and used on the pilots' flight helmet visor housings as early as 1949.

The center decal (Author's collection) and the lower left decal (John Slezak collection) show two different examples from the F-4J "Phantom" years, one with the Naval Air Training Command seal, and the other with U. S. Naval Aviator wings of gold.

The crests didn't start showing up on the back of the flight helmets until the very early 1970s a practice that continues today. As you may have guessed, Protection, Inc. was customizing the Blue Angels' flight helmets when this change occurred. Was this addition the brainchild of the Blue Angels or Protection, Inc?

The answer remains unknown.

21
Don't Judge a Book by its Cover

Collectors are a unique subset of society. Most are driven by the thrill of the hunt and often referred to as obsessive and persnickety. I can't disagree because I have been one of those collectors for the better part of my adult life.

I speak with a reasonable amount of certainty that *all* collectors, regardless of focus or interest, whether it is antiquities, coins, books, paintings or flight gear, will agree that dealing with authenticity is an ever present challenge. For centuries, the collecting world has been ripe with counterfeits, fakes, reproductions and replicas. The purpose of these "imposters" are spawned from many different intentions, some innocent, some not so innocent. Whatever the motive, savvy collectors learn to live by the rule, "Caveat Emptor" — Let the buyer beware.

In the world of military flight gear collecting, and in particular flight helmets, determining the authenticity of a particular specimen is an ongoing challenge with many collectors, new and experienced alike. Sadly, the fledgling collector, without adequate experience or guidance, can easily be enticed into purchasing an item from an unscrupulous or less-than-honest seller, attempting to sell a "knock off" to an unsuspecting buyer as genuine or rare. These are not back alley deals under the cloak of darkness. One needs to look no further than the sometimes dark world of online auctions to find dishonest sellers lurking in the shadows. While the overwhelming majority of sellers are honest, there are a handful that tend to give the rest a black eye.

To be clear, I do not have an issue with sellers who are forthcoming about an item that is not authentic and properly listed as a replica or reproduction. For example, problems begin when a seller lists an item, let's say a flight helmet, as authentic, rare or genuine, when it is clearly a repurposed motorcycle helmet using a combination of genuine flight helmet components and hardware store parts — a shameful misrepresentation at a minimum. To reiterate, the motive or intent is critical to know. If the seller is inexperienced or ignorant of the item's authenticity, that is one case that can often be remedied through off-line, informative communications. However, if the intent is to trick or otherwise defraud the buyer with intentionally deceptive descriptions to realize a greater profit, then these sellers are doing a grave disservice to dedicated collectors who are trying to be honest caretakers and temporary custodians of these aviation artifacts.

The big question is, "How does one determine the authentic from the imposters?" The answer is not easy. First, as with any hobby or interest, educate yourself. There is no substitute for knowledge and a healthy appetite to learn is essential for anyone wishing to become competent in their field of interest. For example, I have studied Blue Angels flight helmets for the better part of 25 years. While I have attained a respectful level of competence on this matter, I *do not* consider myself an expert. I want to be clear: I AM NOT AN EXPERT.

So, how does one educate oneself on such an obscure topic? Again, this is not an easy undertaking. Further complicating this effort is the lack of readily available reference manuals and books from which to draw basic information. Instead, it is important to harvest institutional knowledge from those people who build, customize and maintain these beautiful — and functional — flight helmets. Other potential sources include museum collections and to some degree, the end users: pilots. So, who are these people and what can be learned from their unique areas of expertise?

GENTEX Corporation

The General Textile Mills - now doing business as GENTEX - is a privately held company that focuses on the manufacture of Personal Protective Equipment products for the U.S. and international military, special forces, commercial, law enforcement, emergency medical services and First Responders.

GENTEX produced its first hard shell helmet in 1948. The helmet, known as the H-1, was made for the U.S. Navy and the P-1 soon followed for the United States Air Force. These technologies were quickly incorporated into helicopter pilot helmets for both military and commercial applications, as well as tank crew helmets.

In 1969, GENTEX bought Protection, Inc., a helmet manufacturer in Pomona, California. Ten years later, Protection Inc. began to acquire equipment in order to manufacture respiratory products. By 1981, the new branch officially transitioned into GENTEX Western Operations focusing on the manufacture of respiratory products. The California operation moved to Rancho Cucamonga in 1992.

I understand that the GENTEX plant is not open to field trips and public tours, but employee friendships can be cultivated and are excellent sources of information — trust me, I've done this and it paid great dividends. I would also highly recommend keeping an eye out on eBay for vintage photos, letters, documents, sales literature and publications that almost without fail provide valuable insight into the company. Many of these items pop up in a key-word search that includes "media," "VIP" or "press kits."

Likewise, there are collectors across the globe who have similar artifacts in their private collections. I know some who have been willing to share their data (scans or photo copies) or sell it to you if the time is right. Bottom line is, if you desire knowledge, you must go out and look for it, as it rarely falls into your lap!

I highly recommend networking with like-minded collectors, participating in the exchange of information on FaceBook pages dedicated to flight gear collecting, conducting online research and visiting aviation museums — many of them offer free archive research.

In 1983, GENTEX sold the custom-fit helmet business, doing business as Protection, Inc. to Flight Suits, Ltd., currently doing business as Gibson & Barnes.

Gibson & Barnes

Gibson & Barnes, previously doing business as Flight Suits, Ltd. (FSL), is a manufacturer of flight suits, uniforms, jackets, leather jackets, flying helmets and communications equipment. Customers are pilots, law enforcement officers, emergency medical personnel, military, test pilots, airline pilots and individuals.

United States Navy Blue Angels Flight Helmets

Gibson & Barnes started manufacturing with two employees in a 1,000 square foot building in 1977. Over the years, slowly adding products and customers, they operate a 35,000 square foot facility and employ 115 people. Sales include products to more than 60,000 customers in the United States, Canada, and over 60 other countries.

Gibson & Barnes customers include the all branches of the U. S. armed forces, police helicopter units, hospital helicopter programs and NASA astronauts. They have also painted the US Navy Blue Angels and the USAF Thunderbirds flight helmets since 1984.

In the mid-1990s I was fortunate to develop a business relationship with FSL and, in particular, the Blue Angels Account Manager. I gained a tremendous amount of information regarding the work performed by FSL, specifically the customization of their iconic flight helmets.

In addition to learning the DuPont Imron paint code used to paint these helmets, I was able to acquire several authentic Blue Angels "decals" like the one pictured above. Technically, these were not decals. The images were screen printed on a clear acetate medium with an adhesive backing. The graphics were hand cut into three subcomponents (next page) that helped to make it easier to apply to the compound curve of the helmet shell. Having authentic graphics to compare and contrast with other flight helmets provides an invaluable reference.

It is also worth noting that even though these graphics are professionally printed some imperfections are encountered and are not to be be confused with a reproduction or knockoff.

The Naval Air Training Command "disc" (bottom left) is from the flight helmet of Mike "Nordo" Nord, who flew with the Blue Angels in 1980. These graphics are screen printed in layers from the lightest ink first, followed by progressively darker inks. The white would have been printed first, followed by the yellow and finished with the dark blue. Alignment of the individual screens is critical so the images align with each other, and is more difficult with small jobs such as this 1 inch disk. This particular printing run experienced some alignment issues, referred to as register slippage. In this example, the yellow and white screens were misaligned causing the yellow to encroach into the white and the white to encroach into the yellow.

The disc on the right is from a more contemporary (F/A-18 era) flight helmet and is of overall better quality but, not what I would consider a sharp graphic.

Also beware that the Blue Angels crest decals — as used on the backs of their flight helmets — were produced in two different versions. One version — that featured the NAVAL AIR TRAINING COMMAND "disk" — with the U. S. Navy wings of gold on the bottom, is the proper one used on the rear of the flight helmets (left).

The second version features a blue circle in place of the Naval Air Training Command disk with a large U. S. Navy wings of gold across the center. The latter style is used for the Blue Angels' flight suits and ground crew uniform crests (right).

Provided we are lucky enough to handle an authentic Blue Angels flight helmet, there are a few things that should be examined provided the owner or caretaker has provided explicit approval and careful oversight. Underneath the visor housing are some unique markings hand written by the person responsible for customizing the helmet, whether it be Protection, Inc., Flight Suits, Ltd., or Gibson and Barnes.

Protection, Inc. often applied an adhesive paper tag that included the customer name or number, order number, helmet size, as well as paint and graphics instructions. There are also examples — particularly with Flight Suits, Ltd. and Gibson & Barnes — of this information being written in black marker directly onto the inside of the visor housing.

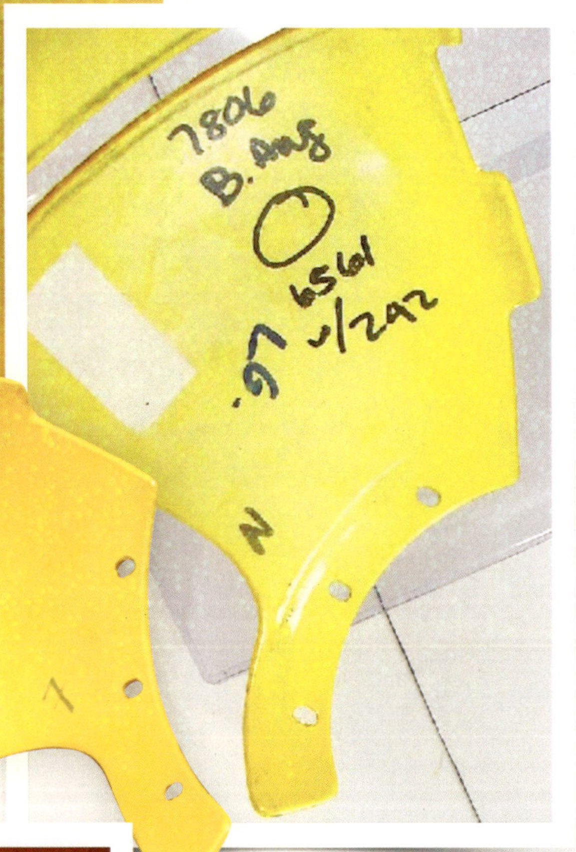

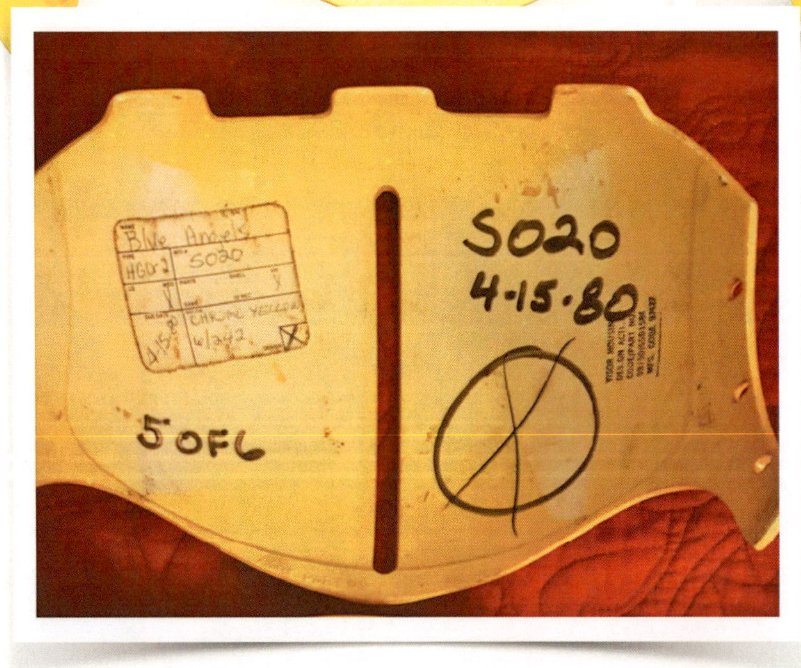

Included on this page are examples of these "instructions" that tell an interesting story. Blue Angels was often written as "BA" along with two 4-digit codes, one being the last 4 digits of the pilot's social security number and the other — 6561 — that referred to the Dupont

Imron Paint code for Blue Angels yellow. On 2 of the examples (previous page), the numbers "2" and "7" refer to the aircraft / pilot number as it applies to the relative aircraft assignment.

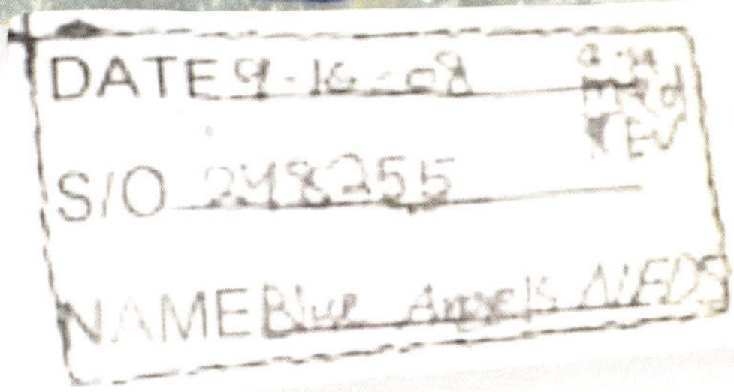

There's more. The custom leather liners inside of the Blue Angels flight helmets can easily be removed for cleaning, repair or replacement. By removing the liner, additional information is often located on the inside of the helmet shell.

The tag on the top right is an early example of an adhesive backed foil tag common to the period listed. This one — as discussed in Chapter 8 — is from Dick Schram's APH-6 flight helmet when he flew with the Blue Angels from 1969 to 1971.

A more contemporary example (above right) from 2008 is from a Blue Angels helmet customized by Flight Suits, Ltd. (now doing business as Gibson & Barnes). This fabric tag was sewn to the Tempo lining of the shell. This example indicates the customer (NAME) was "Blue Angels NFDS." The letters stand for Navy Flight Demonstration Squadron.

It is also prudent to examine the manufacturer's name located on the oxygen mask bayonet receiver "diamond" plates. They are typically branded Protection, Inc., Flight

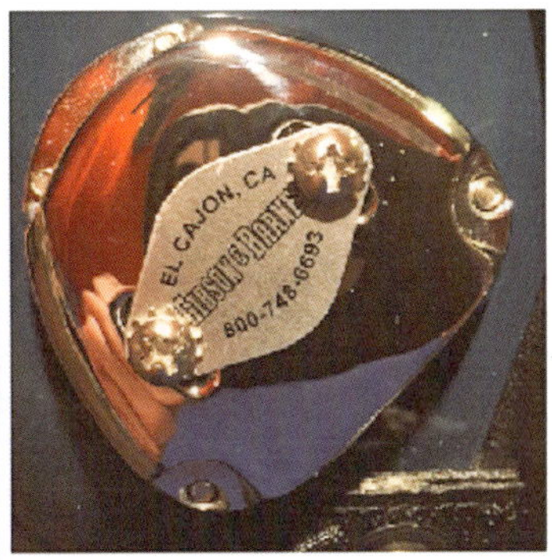

Suits, Ltd. (2 versions) or Gibson & Barnes. Keep in mind that these do not guarantee authenticity, as they can easily be transferred from one helmet to another.

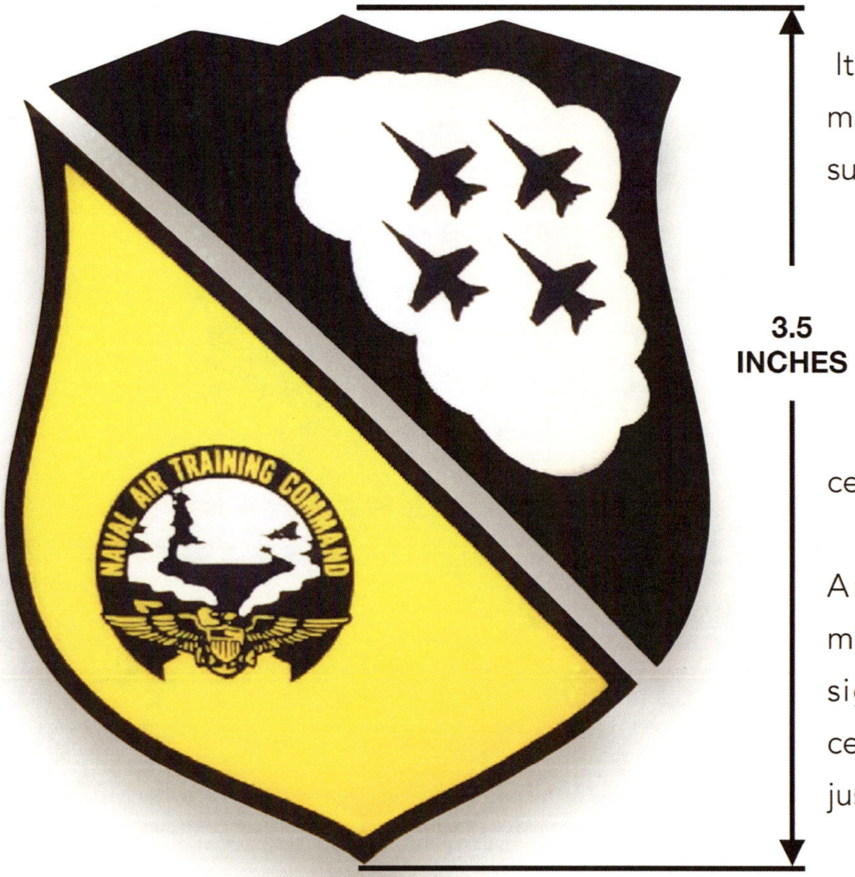

It is important to know as many details as possible, such as the precise size of a contemporary Blue Angels flight helmet decal / graphic. The crest measures 3 1/2" in height, or 8.89 centimeters.

Another detail to be mindful of is the pilot's call sign that is positioned centered below the crest, just above the leather edge roll. The monikers are the same color as the blue field on the crest and used a font originally called "Dom Casual" but is now simply called "Dom."

This is an example of "Dom" font.

Blue Angels Life Support Shop

When seeking a firm grasp or mastery of flight gear components, it is hard to beat the institutional knowledge of a military life support shop. The men and women found here are highly trained, detail oriented specialists that are second to none when it comes to determining the authenticity of an item used in life support. The Blue Angels are no exception. The trick is to establish a meaningful relationship with someone in this career field, which is easier said than done. I will share why.

The life support specialists with the Blue Angels rotate in and out every 2 to 3 years, so forget about a long-term relationship while serving with the Team, provided you can establish one. Introductions can be made via phone and / or email, but at some point an in-person meeting is critical to build a professional and meaningful relationship. This can be accomplished by meeting at a scheduled airshow, or at their squadron located at NSA Pensacola. Finally, patience is key. These folks spend more time on the road than at home, so they can become elusive, moving targets!

I was fortunate to make my first Blue Angels life support contact in 1997 via a cold call. In my favor was we quickly realized we grew up about 15 miles from each other! That call led to several visits to NAS Pensacola, the Blue Angels life support shop and a friendship that continues today. He also introduced me to another life support specialist while we were both attending the 2018 Blue Angels Homecoming airshow at NAS Pensacola. This led to another visit to the life support shop where I was afforded an opportunity to photograph flight helmets and have some burning questions answered. These introductions are priceless as it provides some validation to who you are and affirmation of your research.

I have been ringing phones, writing letters, sending emails and speaking to anyone who will take the time to speak to me for 25 years and I *still* have questions! The thirst for knowledge should never end. Don't get me wrong, writing a book certainly opened some doors, but the struggle is real!

Blue Angels Flight Helmet "Hall of Shame"

So, were you hoping for a gallery of Blue Angels flight helmets that are worthy of the dreaded "Hall of Shame?" Well, that's not going to happen. Allow me to offer my best explanation.

There are more than enough photos of "questionable" — trust me, I am being polite — Blue Angels flight helmets on the Internet and auction sites that are clearly (in my humble opinion) not authentic. I'm talking about pages of photos. However, I have chosen to take the high road. I realize that such a bizarre collection may be oddly entertaining or perhaps deemed a valuable resource to those who are making a concerted effort to educate themselves. Instead, I have chosen not to "call out" those "bad" examples for fear of bruising some egos or worst, being accused of slander, defamation or character assassination! I have come to terms with the harsh reality that I could be wrong. So, it bears repeating, I AM NOT AN EXPERT. Besides, some sellers / social media posters may not know that what they have is not authentic and I don't feel this is the venue for public shaming. For the purpose of this book, it is best left unsaid.

What I will share with you are a few examples of known "imposters," "fakes," "replicas," call them what you may. A few of my flight helmet collecting colleagues, who will remain anonymous, call them "kitchen jobs" — a term with which I am not particularly fond. Whatever you choose to call them, I am prepared to offer the reader the following examples to compare and contrast with authentic helmets found throughout this book. Not to worry, I have secured the express consent of the creator to post these images and I am confident that there will be no ill feelings. I know this because I created them.

I hope you find the images on the following page to be educational and entertaining.

United States Navy Blue Angels Flight Helmets

In my companion book, *The Red Helmet*,[1] chronicling the history, development and evolution of USAF Thunderbirds flight helmets, I discovered several anomalies worthy of mentioning here.

I had to re-think...are they in fact imposters? I learned that just because it looks like a duck, and sounds like a duck, its not always a duck! Allow me to explain.

With alarming frequency, I was shocked to learn — from the pilots' own words — that their coveted Thunderbirds flight helmets were so comfortable (custom fitted) that they often painted over the iconic red scallops and blue stars when they returned to regular squadron assignments! I can only imagine that following their Air Force careers, or the life expectancy of the flight helmet, these rare beauties, cloaked in mil-spec white paint or woodland camouflage, were unwittingly tossed in the garbage or sent to DRMO[2] to be sold as scrap or worse, destroyed.

In the most egregious example, I had the fortune of tracking down an exceptionally rare, TOPTEX T-3 flight helmet issued to and worn by one of the USAF Thunderbirds demonstration pilots from the 1964 show season. For those who may not be familiar, the Thunderbirds were flying the massive F-105B "Thunderchief." After only six shows, the #2 Left Wing jet suffered a major, catastrophic structural failure, which took the life of the pilot.[3]

The helmet I was "bird dogging" belonged to one of the surviving pilots, and rumor had it from a close family member that it was on display at a small, private club in Florida. Through a series of significant contacts (who will remain anonymous), I verified that the helmet was in fact on display, and my request for detailed photos was granted.

---

1 Elm Grove Publishing, ISBN 978-1-943492-76-3

2 Defense Reutilization and Marketing Office (DRMO) changed name to DLA Disposition Services. The site now is called DLA Disposition Services. Military Property Turn-In. DLA Disposition Services is responsible for the disposal of excess DOD personal property, foreign excess, personal property (FEPP), scrap, hazardous waste, and property requiring demilitarization.

3 Capt. Eugene J. Devlin, 09 May 1964, Hamilton Air Force Base, California

The photos were promptly emailed to me. I am having difficulty finding the words to describe my horror of what I had witnessed. The flight helmet was in fact a TOPTEX T-3 as expected for this time period. However, there were several anomalies that had me perplexed.

The classic, nickel plated, Hardman Tool & Engineering Company oxygen mask receivers (the same ones used on the Blue Angels APH-5 flight helmets - see Chapter 7), had been removed. In their place, were contemporary, lightweight oxygen mask receivers like the ones used on modern HGU-55/P flight helmets (see Chapter 11). To add insult to injury, the helmet was fitted with the wrong oxygen mask, but this was only a minor infraction in a much larger, travesty of epic proportions: The beautiful, buckskin leather edge roll — iconic to these period flight helmets — had been dyed or painted black! As if that were not bad enough, the red scallops, blue stars and Thunderbirds emblem had been…wait for it…REPAINTED!

Following some discreet and gentle inquiries to the person sending me the photos, I was told that the pilot (who is now deceased), had reached out to his old crew chief from the 1960s and commissioned him to "refurbish" his flight helmet to reflect those used by his Thunderbirds brothers and sisters flying the F-16 "Fighting Falcon!" And yes, the Thunderbirds emblem prominently displayed in the rear center of the helmet was a contemporary version depicting a four-ship diamond formation of…F-16s! It continues to haunt me as I attempt to understand why anyone would want to do something such as this. Perhaps I should let this go, but it's easier said than done. I guess I am driven by a sense of "responsibility" to preserve history, not to mask or destroy it.

But wait, there's more.

This story comes to me by way of a friend, "brother in blue," historian and fellow collector, David Mattsson. Dave tells me he had a friend[4] who owned and operated a flight simulator store in Minnesota. They shared a passion for all things aviation.

---

[4] Mike Pohl. Sadly, now deceased.

One day, while visiting Mike at the shop, he learned that Rick Adams, Blue Angels #7 - Narrator for the 1968 and 1969 show seasons was invited to make a scheduled appearance. Mike encouraged Dave to enter a drawing for a once-in-a-lifetime opportunity to fly simulated formation in an F/A-18 simulator...with a former Blue Angel! "Lady Luck" smiled on Dave as his name was drawn and announced. Dave was introduced to LT Rick Adams (1968 - 1969). Dave said, "So for a half hour, I was able to fly around with Rick, chasing each other through the virtual sky, flying formation, and just having fun, flying with a Blue Angel! I'll never forget that unexpected treat!"

Following the simulated flight, Rick Adams entertained the store crowd with stories of his flying days in Vietnam and the Blue Angels. As the conversation wound down, Dave stared in awe as Adams presented a well used, APH-6 flight helmet, sporting VF-11 "Red Rippers" reflective tape, to be donated to Mike. Dave said he almost soiled himself as he watched with envy the rare transaction unfolding before his hungry eyes!

"So for a half hour, I was able to fly around with Rick, chasing each other through the virtual sky, flying formation, and just having fun, flying with a Blue Angel! I'll never forget that unexpected treat!"

David Mattsson

Unfortunately, Mike passed away. Dave was successful at purchasing the "Red Rippers" helmet from the estate, where it resides today in his significant flight gear collection.

As most flight helmet collectors know — and Dave is certainly no exception — there are often fascinating stories waiting to be told if you are willing to look beneath the surface under the many layers hidden by the passing of time. Dave did just that and what he found was most satisfying, to say the least!

After carefully removing the fragile, custom liner away from the shell, the liner was clearly marked "ADAMS" and "L APH-6" which likely refers to a LARGE APH-6 liner. Even more interesting were the markings on the shell under the liner: "ROTTGERING" written beneath "ADAMS." For those who may not be familiar, Lt. Dave Rottgering was LT Rick Adams' predecessor as the #7-Narrator for the 1966 and 1967 show seasons. Could this shell have been first issued to and worn by LT Rottgering, then assigned to LT Adams before finding its way back into the fleet in VF-11? Anything is *possible*. But, the story doesn't end here.

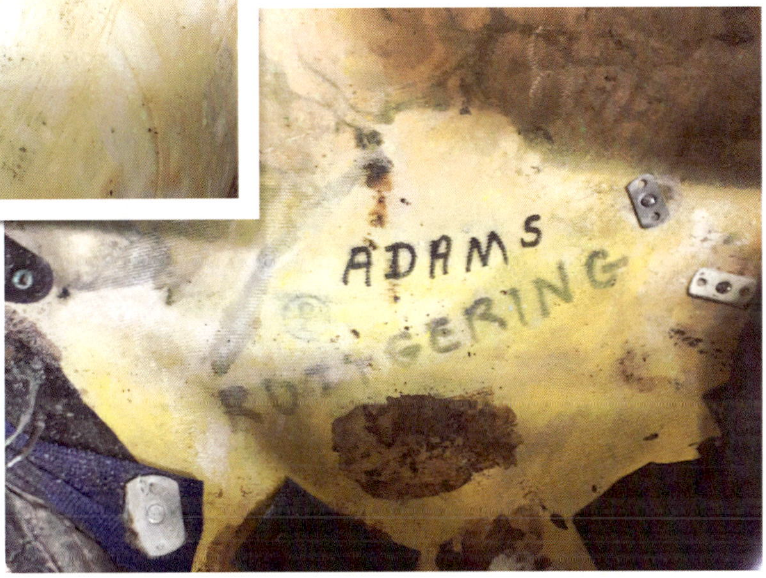

Although the reflective tape that represented the "Red Rippers" of VF-11 covered the entirety of the helmet shell, small gaps are not uncommon between tape strips as well as where the edge roll is attached to the shell. Upon closer examination, David determined that prior to the helmet receiving the custom tape job, it was originally painted…wait for it…Blue Angels yellow!

There are many people — I'm one of them — that would want to painstakingly remove that layer of reflective tape to reveal a rare, vintage, stunningly beautiful, historically significant Blue Angels flight helmet…named to not one, but two pilots! Dave admits he is curious but, resists the urge based on the following sound reasoning.

First and foremost, the helmet as it exists, is a beautiful, squadron-taped VF-11 "Red Rippers" flight helmet that represents an actual piece of surviving history and deserves honor, respect and preservation. I can't disagree.

Secondly, there are at least two possible outcomes to removing the reflective tape, neither of which are good. The first — and most obvious — is that the unique identity of this VF-11 "Red Rippers" helmet would be destroyed and lost forever. Most would agree that this is not an acceptable option.

Let's assume you conduct a risk analysis and the decision is made to remove the tape in the hope that what's underneath is of greater historical significance than what meets the eye. I am here to tell you that referring to this procedure as a "crap shoot" is a gross understatement. If you have never removed reflective tape from a helmet shell, then allow me to explain the process in excruciating detail.

Removal of reflective tape — not to mention tape that's 50+ years old — can be challenging. The longer tape has been applied to a helmet, coupled with decades of sun exposure and temperature swings, the more stubborn it is to remove. Many techniques exist, and everyone has their favorite. Some use mild to medium heat to soften the adhesive, while others use chemicals, like acetone or lighter fluid, the latter of which is a butane or naphtha based product. Whichever technique you favor, one thing is certain, it is messy, the glue is nasty, the process can be labor-intensive, and the results can vary...*greatly!*

If we apply this process to the "suspected" Blue Angels livery cloaked in "Red Ripper" clothing, we need to be cognizant of the following potential disasters. While the use of heat will aid in the removal of the tape, it could also soften the paint and / or clear coat making it susceptible to bubbling, damage or lifting. Similarly, if chemical aids are used, there is an inherent risk that there may be an adverse reaction to the paint and / or clear coat underneath resulting in irreparable damage.

Just for argument sake, there may be those who suggest no heat or chemicals. This is possible but, get ready for the long haul. This technique is the safest but, takes time and patience, because the tape is typically very brittle and comes off in tiny shards as opposed to long strips. If you're lucky, some older tapes sometimes separate easily from the adhesive, but the glue that's left behind is hard, dry, crusty and difficult to remove. It is almost impossible to remove reflective tape from a previously painted surface without some sort of damage, whether it's dents, nicks or worse, the tape lifting the paint from the shell.

This discussion would not be complete without talking about the worst case scenario. Hypothetically, the decision is made to remove the tape, which means the VF-11 helmet ceases to exist — BAD. The tape is successfully removed to reveal a plain yellow helmet with no Blue Angels graphics — BAD. The tape is removed to reveal a vintage Blue Angels flight helmet but, the process has taken a terrible toll on the paint and graphics — BAD.

At the end of the day, Dave is the sole custodian — and final authority — when it comes to this aviation artifact and has made the executive decision that the risk outweighs the reward. This beautiful helmet has revealed many secrets, but harbors a few more. I, for one, respect this decision.

So, what does this all mean? My point is to reiterate how this chapter began, "Caveat Emptor" — Let the buyer beware. While there's no shortage of "imposters" shamelessly circulating in and out of online auctions and personal collections across the globe, we also need to pause, contemplate and be cognizant of the possibility that there could be something of much greater value underneath what meets the eye, disguised in plain sight as another squadron...or a "kitchen job!"

Remember the old axiom, "Don't judge a book by its cover."

VF-11

APH-6 flight helmet from VF-11 "Red Rippers" named to Dave Adams.
(Private collection of David Mattsson)

The photos on this page clearly illustrate that this VF-11 "Red Rippers" helmet was previously painted a very familiar yellow.

The circled area further shows that the yellow paint includes a fine, silver metal flake, a telltale trademark of Protection, Inc. as well as the Blue Angels' flight helmets they produced during this era.

(All photos courtesy of and from the private collection of David Mattsson)

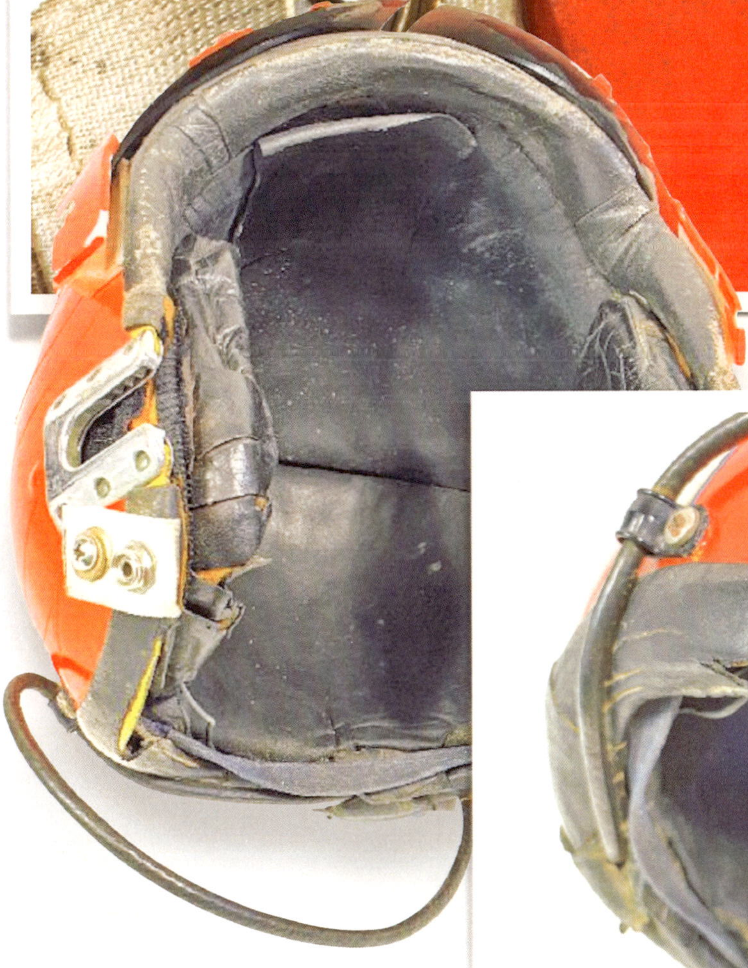

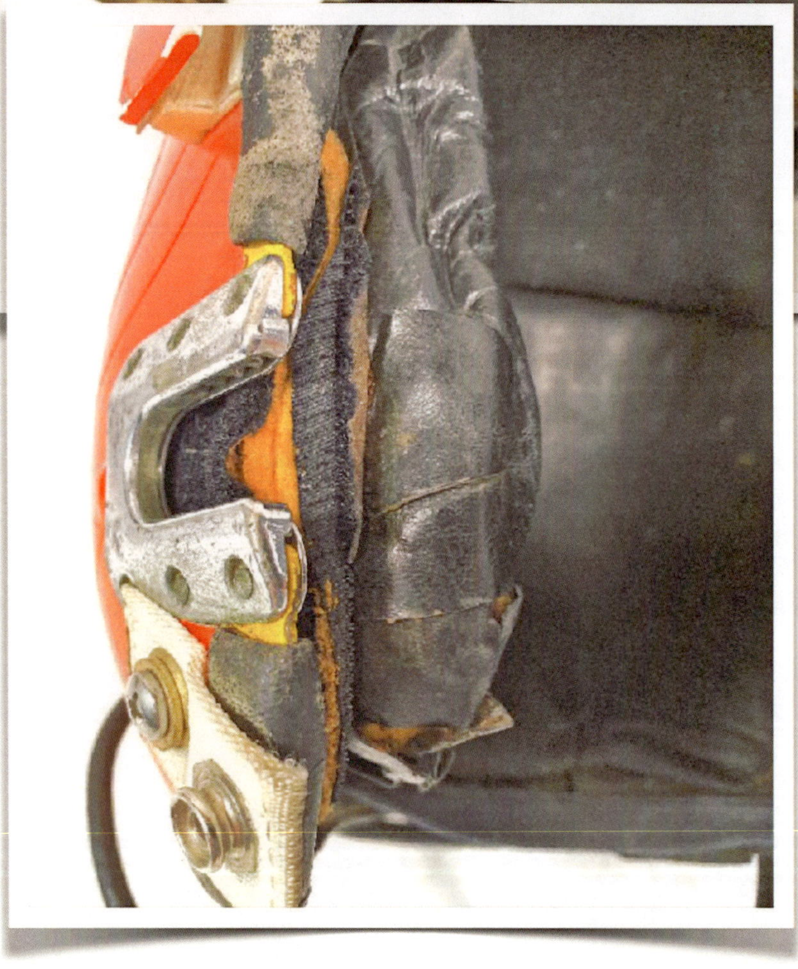

Additional photos showing that the yellow paint under this tape covers the entirety of the helmet. Inquiring minds want to know but, not at the expense of destroying the VF-11 "Red Rippers" history, or what may hide underneath.

(All photos courtesy of and from the private collection of David Mattsson)

I continue to be intrigued and fascinated by these stories and can only begin to speculate on how many more of these examples are out there hiding in plain sight. All the more reason to take the collection and preservation of these unique and often rare artifacts seriously.

We often think of the National Museum of Naval Aviation at NAS Pensacola or the National Museum of the U. S. Air Force at Wright-Patterson AFB for their unparalleled collections of rare and historic aviation artifacts. And, I would not disagree. However, I would venture to say that in the case of military flight helmets, the vast majority reside in the collective hands of private collections. I have witnessed this first-hand in my book research. Fortunately, throughout history, countless precious artifacts have been rescued from garages, attics and flea markets by individual, conscientious collectors. Unfortunately, most of these private collections remain private and are not shared with the public.

As a collector and historian myself, I salute all of the aviation museums and private collectors alike, for their tireless efforts towards preserving our rich aviation history.

Flight Gear Collectors patch designed by fellow Belgium collector, Niki Helsen.

22
The Blue Suit

This book represents a journey. A life-long journey fueled by an ever-burning passion for aviation. This journey has no destination only stops along the way where likeminded people meet, old stories are shared and new memories are made. Some of the stops on our journey are intentional, others unplanned, but all contribute equally to a story that is uniquely ours and like no other.

My "Blue Angels journey" began in 1974 and continues today. The people I have met along the way have been instrumental — some directly, some indirectly — in my research for this book. Although the subject matter is quite specific, I purposely cast a very broad "research net" for two main reasons: 1.) I have an interest in all things Blue Angels, not just flight helmets, and 2.) I have learned from experience that you often discover information in places, and with people, where you least expect to find it. Something I learned from a 35 year career as a professional investigator.

This journey, or this book, would not be complete without mentioning a small tailor shop on North Navy Boulevard in Pensacola, FL called The Sewing Box (formerly doing business as "Feminine Flair"). The unassuming storefront sits among a few other small businesses in a strip shopping center. Even with a physical address, it's not always recognized by Google Maps, and should you be lucky enough to make it to the parking lot, it still requires a keen eye to spot the understated sign above the front door.

Inside the shop is, well, much like any other tailor shop, crowded, the buzz of commercial sewing machines fill the air, prom dresses and military uniforms in various degrees of completion are hanging silently and patiently as they await final fittings.

The man who started the business known as Feminine Flair, Inc. is Samuel Miller, a U. S. Navy Master Chief cryptologist, who retired in 1976. Sam always had a passion for designing women's clothing, and it is from this dream that a second career was launched.

Sam capitalized on his previous military career, and location near NAS Pensacola, by custom tailoring women's military uniforms, a much needed service as the Navy was challenged with the increasing number of females throughout their ranks, and difficulty meeting their specific clothing requirements.

In 1986, Sam was awarded a contract to make the iconic "Blue Suit" for the U. S. Navy Blue Angels, and continues to make them to this day. These flight suits as presently constructed were personally adapted by Sam from the original design. Each one is hand-made and customized to fit each Team member like a glove. First year Team members receive 6 suits, and returning (2nd year) Team members receive 4 suits.

Sam sold the business in 2013, and the shop name was changed from Feminine Flair to The Sewing Box; however, he continued to work part-time tailoring the Blue Angels flight suits. In 2019, the business was sold again, this time, to long-term employee, Candy Whitehurst. Sam's expertise and institutional knowledge was again retained.

Over the years, Sam has become an advocate for my research and military aviation history preservation efforts. He has introduced me to numerous Blue Angel pilots, contributed memorabilia to my aviation museum, and most importantly, become a great friend. The photo above is of the author (left) and Sam Miller (right) during a visit to Pensacola for the 2016 Blue Angels Homecoming Air Show. The inset photo was hanging on the front door, proudly advertising 30 years of support to the Blue Angels.

(U. S.Navy photo)

Some examples of the stunning blue and yellow flight suits custom tailored by Sam and Candy at The Sewing Box. These iconic flight suits were designed by Sam and have remained virtually unchanged for 30+ years, standing the test of time. As the saying goes, "If it ain't broke, don't fix it!"

Well done, Sam!

Photo credit:David Colwell
Baltimore Magazine

On this page are several examples of the embroidered name tags and wings that are sewn onto the Blue Angels' flight suits. (Author's private collection).

In 2015, Sam was visiting family not far from where I live, and I was fortunate to invite him into my home as a guest. We shared many Blue Angel stories, some of which I agreed not to share publicly!

In the photo on the left, Sam is proudly standing next to two Blue Angel uniforms in my personal collection and on display in my private museum, the Bagdad Aero Club.

Sam confirmed that he in fact made both of these flight suits and remembers the pilots. When I acquired the khaki C-130 flight suit, it was "de-milled" by cutting the Blue Angel crest out of the right breast. I contracted Sam's tailoring skills and he returned the suit to "like new" condition with fabric still on hand in the shop.

Sam has openly shared that one of his goals before retiring as "Tailor to the Blue Angels" is to measure and make a suit for a "numbered" female pilot. "Numbered" pilots are the ones who fly the six-ship show aircraft: 1 - Leader / Boss; 2 - Right Wing;

3 - Left Wing; 4 - Slot; 5 - Lead Solo; 6 - Opposing Solo; 7 - Narrator, and 8 - Events Coordinator.

The Blue Angels had a female C-130 "Fat Albert" pilot, but are yet to have a successful candidate for a numbered jet position.

Sam was interviewed in 2015 for *Splash Magazine* and shared that once Blue Angel candidates reach the final round of the selection process (usually around the 4th of July), Sam's shop makes the trip to the Blue Angels' squadron to measure the final candidates for their "Blue Suits." Unfortunately, not all are selected, and a few never return. A harsh reality, and only one of several "gut checks" along the way to becoming a Blue Angel. Those selected for the Team will make subsequent trips to Sam's shop for the initial and all subsequent fittings.

Recently, Sam thought his wish would come through. A female pilot came into his shop as part of the group of final candidates. Sam was quoted, "There was a female pilot who made it to the last round for the F-18s. Before I took her measurements, I told her, 'I've been waiting for you for 29 years, and if you don't get picked, I'm writing a letter to the Secretary of the Navy.' She didn't get picked, but I sure hope she comes back."

In 2019, Sam granted an interview to the *Pensacola News Journal,* where he was quoted as saying he will give up his rotary cutter and measuring tape once Cdr. Brian Kesselring, the man who will lead the Blue Angels through the 2020 and 2021 seasons, gives final approval to the suit Miller is now creating for him.

Sam said, "I have arthritis and my fingers don't work like they should. I don't want the quality of the work to slip."

Recently, the Blue Angels changed their contract status with The Sewing Box from "sole source provider" to open bid, but it is unlikely there will be any changes in the near future. Candy Whitehurst was quoted, "I think it's looking good. I don't think anyone else can really do what we do." I totally agree, Candy!

The photo on the right is of Sam with CAPT Tom Frosch (Boss, 2013 to 2015), as he is presented with a custom flight log book cover. In recent years, Sam and Candy have begun making sofa pillows from flight suits and ground crew uniform shirts for a one-of-a-kind memento of a Team members service with the Blue Angels.

In August 2019, Sam Miller "officially" retired, leaving behind a legacy that will live on forever. His first retirement was as a Master Chief. This retirement is as a Master Tailor. Sam has been a great teacher over the years and leaves the shop in good hands. I have no doubt the Blue Angels will be return customers for many more seasons to come. Best of luck to Candy!

If I were a betting man, I wouldn't be surprised to find Sam at the shop every July to welcome the new Team members, share some stories from the past 3 decades and hoping to measure that elusive first female, numbered jet pilot.

Hang in there Sam, she's out there somewhere, waiting to be discovered.

The photo on the left is the first "Blue Suit" Sam made for Gil Rud, the Blue Angels Commander in 1986. On the right is Sam with CDR Brian Kesselring, the Blue Angels Commander for the 2020 and 2021 show seasons, as he is fitted for his first "Blue Suit" and Sam's last...or so, he says!

In closing, the "torch" has been passed to Candy Whitehurst, a talented and worthy successor to the "Blue Suit" empire proudly built by Sam Miller. I am confident this legacy will not only survive but, thrive for many more seasons to come as we witness, welcome and celebrate the next generation of "Tailors to the Blue Angels."

Best of luck, Candy and thank you for your support, but more importantly, I am grateful for your friendship!

23
Trader Jon's

If you never had the opportunity to meet Martin Weissman or visit his notorious Pensacola nightclub, then it will be challenging to adequately describe the legendary man better known as "Trader Jon." In a word, eccentric comes to mind but, that would be an unfairly simple description of a far more complicated man. Perhaps I should start at the beginning.

Martin "Trader Jon" Weissman (1915 - 2000) was born in Brooklyn, NY. During training as a U. S. Army paratrooper, he broke his ankle, an injury that left him permanently bowlegged. As a result, he was honorably discharged, but maintained a life-long love for the military, and in particular, military aviation.

In 1953, Trader and his wife moved from South Florida to Pensacola where they opened a bar at 511 South Palafox Street. The namesake bar — Trader Jon's — quickly became a favorite watering hole for U. S. Navy and USMC officers stationed at nearby NAS Pensacola.

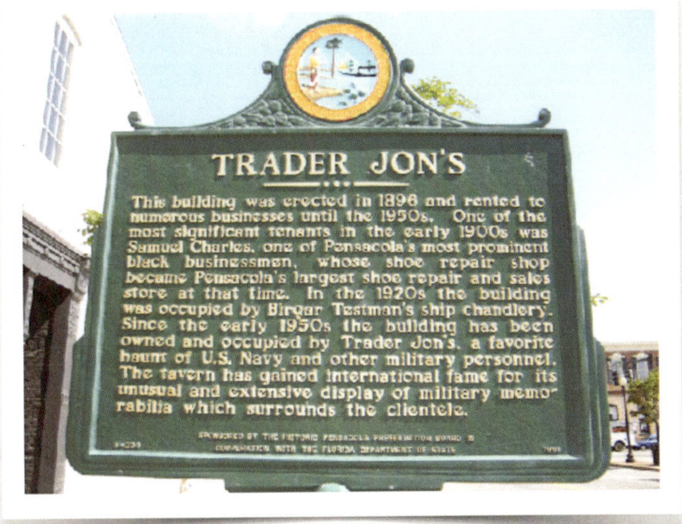

The news travelled fast, and soon pilots from across the country and from all service branches were planning cross country "training" flights to Pensacola to visit the famous — or infamous — Trader Jon's.

The 1982 Hollywood film, *An Officer and a Gentleman*, was based on the Officer Candidate School at NAS Pensacola, and Trader Jon's was the inspiration of the fictional club called "TJ's".

Trader Jon came by his name honestly, often trading drinks for aviation memorabilia which he proudly displayed in the bar. There were also no set prices for drinks, rather he relied on a system known as "Traderomics" which translates to varying prices depending on his mood, or how well he knew you.

Trader Jon's also enjoyed a cult following both within and outside of the aviation community. Some notable guests included Prince Andrew, John Wayne, Elizabeth Taylor, Bob Hope, John Glenn, Henry Fonda, Larry King, John McCain, and Neil Armstrong just to name a few.

Despite the unquestionable "A List" status of his celebrity guests, they took a backseat to his favorite patrons. Without a doubt, the guests he held in the highest regard were the United States Navy Blue Angels! And, The Blue Angels loved Trader Jon!

Trader Jon was so fond of his beloved Blue Angels that he added an annex to the main bar that he designated, "Trader Jon's Blue Angel Museum." It was legendary. In the mid-1990s, I had a few opportunities to patronize Trader Jon's, meet the man himself, and actually traded with Trader Jon! The Blue Angel Museum was like a shrine, a place of honor and respect to all Blue Angels past and present.

I also have it from good sources that some parties at Trader Jon's were epic, even rumored to be ripe with bad judgement and questionable behavior. So I am told.

The hallowed walls were adorned with photos from every Team since their inception in 1946. The ceiling was cloaked with Blue Angel flight suits stapled to the rafters, donated by pilots as gratitude to Trader's loyal friendship and hospitality.

On one of my visits, Trader Jon told me that every year, following the Blue Angels Homecoming Air Show[1] at NAS Pensacola, the Blue Angel pilots, ground crews and families held an "After Party" at his bar. To show Trader Jon their genuine appreciation, many pilots would shed their show-flown flight suits and staple them to the ceiling as a small gesture or token of appreciation.

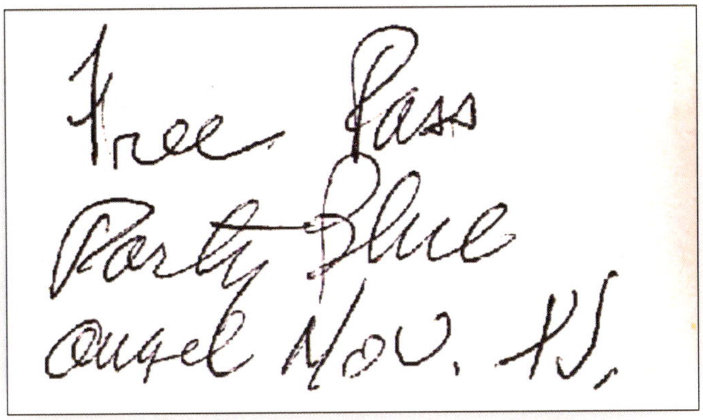

In 1996, my family and I attended the Blue Angels Homecoming Air Show in November, one of many in my lifetime. While there, I was fortunate to have an opportunity to trade an artifact in my personal collection in exchange for a Blue Angel flight suit from the ceiling of Trader Jon's!

Following the "trade," Trader Jon handed me a business card (pictured above), inviting me to the Blue Angels After Party! He wrote on the back of the card, "Free Pass. Party Blue Angel. Nov. TJ." My fortune soon turned to reality as I remembered I had to be an adult — I had my wife and four kids in tow, and we had to make the 5-hour trip home for school and work the following day. Thanks for the invite, Trader…maybe next year.

By the way, Trader Jon is one of only 10 "Honorary Flight Leaders," a title bestowed upon him by the Blue Angels in honor of this life-long support, respect and unconditional love for Naval aviation.

Tragedy struck in 1997 when Trader suffered a stroke, resulting in the closing of the infamous bar in 1998. Trader never returned to work.

---

1 The Homecoming Air Show is the Blue Angels' final show of each season.

When Trader Jon passed away in 2000 at the age of 84, an AP article described his "ramshackle saloon" as a "fixture for aviators from NAS Pensacola." Vice Admiral John H. "Jack" Fetterman was quoted, "The impish Weissman treated everyone with equal affection, from raw recruits to war heroes, astronauts and celebrities." Fetterman added, "He was a loving guy, caring guy who never said anything bad about anybody. You talk about bonding, and you talk about brotherhood, and you talk about what naval aviation was all about. Trader kind of provided that foundation. In my heart, I can't think of that bar without Trader."

Hoping to keep the bar open and the Trader Jon legacy alive, former Blue Angel leader Bob Stumpf and Marine Major Kent Bolin formed the "Trader Jon's Preservation Squadron" and raised $300,000 to purchase the property. The Weissman family rejected the offer.

"Trader Jon's was a place where Naval Aviators went to relax, swap stories and enjoy each other's company. I cherish the memories of my time spent there."

U. S. Senator John McCain

A subsequent offer of $465,000 from a U. S. Navy flight instructor by the name of Matt Heckemeyer and his wife was made, and accepted. After an additional $150,000 in renovations, the bar reopened in September of 2000. Unfortunately, the business never rebounded following Trader's death, and the bar closed for the last time in November of 2003.

The building was once again sold, and the contents purchased by the law firm Aylstock, Witkin & Sasser, and donated to the Naval Aviation Museum Foundation. The memorabilia inside was appraised at about $2 million.

The photo collage on the next page was taken by the author at Trader Jon's on a visit in 1996 or 1997. These are but a few of the many Blue Angel flight suits — spanning many seasons — stapled to the ceiling, an homage to the Team he loved. It was truly a sight to behold. My only regret is that I didn't take more photos.

2 Maj
USMC

Bob Hope

1 Cdr
Bob Stumpf

Cdr
1 Bill Wheat

4 Lcdr

BLUE ANGELS
4 Lt
Doug Thompson

The man…the legend…Martin "Trader Jon" Weissman. I took this photo of Trader in 1996, and had it autographed when I made a return visit.

The Naval Aviation Museum Foundation eventually transferred the artifacts to the University of West Florida, which paid tribute to Trader Jon by opening a display in 2016 in the T. T. Wentworth, Jr. Museum, Pensacola, FL. The artifacts were scheduled to be on display for 3 years; however, the response was so positive, the display will remain indefinitely. The Wentworth exhibit features 15 percent of the estimated 10,000 pieces of military memorabilia that make up the Trader Jon collection.

"This exhibit is unique in that it will resonate not only with the local community, but also with those service members who frequented Trader Jon's while they were stationed in Pensacola and are now spread across the globe," said Dr. Brendan Kelly, vice president for University Advancement and CEO of the Historic Trust. "We wanted to give those who had the pleasure of visiting the real Trader Jon's an opportunity to revisit history and allow those who didn't to experience it for the first time."

Author's photo, Trader Jon exhibit, University of West Florida Historic Trust.

Having visited both the original Trader Jon's and the Wentworth exhibit, I can say without reservation that the effort to preserve the Trader Jon legacy is exceptionally well done. The following pages represent a few photos I took from a couple of visits to the Wentworth Museum. Enjoy!

Author's photos, Trader Jon exhibit, University of West Florida Historic Trust.

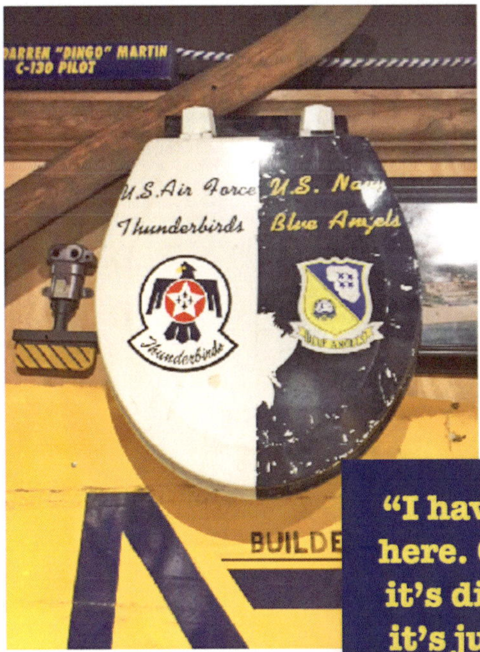

"I have my own world here. Outside the door, it's different. In here, it's just...well, good."

Trader Jon
Trader Jon's - Cradle of U.S. Naval Aviation
James L. Dickerson, Author

Author's photos, Trader Jon exhibit, University of West Florida Historic Trust.

Author's photos, Trader Jon exhibit, University of West Florida Historic Trust.

Trader Jon proudly pointing to the CAPT Greg Wooldridge jacket I traded him for a Blue Angels flight suit. (Author's photo inside of Trader Jon's Bar)

In Memoriam

USN Blue Angels - Pilot Fatalities

29 September 1946 — LT Ross "Robby" Robinson
07 July 1952 — LT Bud Wood
14 October 1958 — CDR Robert Nicholls Glasgow
14 June 1960 — LT Anton Campanella
15 March 1964 — LT George L. Neale
2 September 1966 — LCDR Dick Oliver
1 February 1967 — LT Frank Gallagher
18 February 1967 — CAPT Ronald Thompson
14 January 1968 — LT Bill Worley
14 February 1972 — LT Larry Watters
26 July 1973 — LCDR Skip Umstead
26 July 1973 — CAPT Mike Murphy
22 February 1977 — LT Nile Kraft
8 November 1978 — LT Michael Curtin
22 February 1982 — LCDR Stu Powrie
13 July 1985 — LCDR Mike Gershon
28 October 1999 — LCDR Kieron O'Connor
28 October 1999 — LT Kevin Colling
21 April 2007 — LCDR Kevin J. Davis
2 June 2016 — CAPT Jeff "Kooch" Kuss

Acknowledgements

It would have been impossible to take on this monumental task of research, documentation and editing without the support, encouragement and patience of many people. I would like to thank my wife, Renée for her patience and support for those long hours of research, road trips and endless hours on the computer. I would like to thank my editor, Ed Roy and my publisher, Elm Grove Publishing for their continued support and faith in my work. I would also like to thank the following people for their selfless time and talent by providing me with interviews, facilitating tours and sharing technical data. Without them, this book would not have been possible.

United States Navy Blue Angels Public Affairs
U. S. Navy Office of Information - EAST
Dina G. Linn, Museum Curator, National Naval Aviation Museum
Becky Conrad, Assistant Registrar, San Diego Air & Space Museum
Kevin Drewelow, Director, Combat Air Museum, Topeka, Kansas
T. T. Wentworth Museum, Pensacola, Florida
James Wegge, Director, Gibson & Barnes
Mark Jones, GENTEX Corporation
CAPT Greg "Rugdance" Wooldridge, USN Blue Angels (Retired)
CDR John "Gucci" Foley, USN Blue Angels (Retired)
Lt. Col. Kevin O'Mara, USMC (Retired), USN Blue Angels
LCDR Rick "Timber" Young, USN Blue Angels (Retired)
LCDR Megan M. Shutka
Major Jeff "D.A.R.Y.L." Mullins, USMC, USN Blue Angels
WO3 Brian DeRidder, USN Blue Angels (Retired)
PR1 Travis J. Bradley, Life Support, U. S. Navy Blue Angels
AME1 (AW) Jack Reynolds, USN Blue Angels
PR1 (AW / SW) Darby Versluis, USN Blue Angels
AME1 (AW) A. J. Acosta, USN Blue Angels
AW1 (NAC / AW) Michael H. Johnson, USN (Retired)
PR1 Robert Fugate, USN Blue Angels (1985 - 1988)
MC2 Cody Hendrix, USN Blue Angels
PR2 (AW / SW) C. J. Woods, USN Blue Angels
PH3 Jim Preston, U.S. Navy Blue Angels (1974 - 1977); Supervising Photographer for the Smithsonian Institution, National Air & Space Museum
Bob Moore, USN Blue Angels Aviation Structural Mechanic (1952 - 1953)
Richard Keane, Blue Angels Crew Chief (1969 - 1971)
John Slezak
Steve Norris
Perry Hunter
Sam Miller
Candy Whitehurst

Index

A-4, 7, 10, 16, 129, 149-150, 199, 225, 228, 244
Acosta, A. J., 261
Air Navy Magazine, 110
Aircrew Survival Equipmentman, 96, 218-219, 225, 227, 258
AN-6540, 26-28, 31
AN-6542, 26-27
Anderson, Scott, 156
APH-5, 60, 74-75, 77, 81-94
APH-6, 86, 95-98, 100-101, 104-110, 112, 116, 136, 141, 143
Armstrong, Neil, 316
Ashford General Hospital, 22
Aviation Structural Mechanic, 10-11, 219-220
B-8, 28
Bagdad Aero Club, 310
Bagwell, Bruce, 54, 63
Baltimore Magazine, 308
Band Camp, 168
Bartlett, Russ, 175, 194, 212
Baxter, 167
Beardsley, Bill, 121, 130
Beef, 175, 212
Bell, Gerald M., 57
bellsaviation.com, 57
Bernacchi, Ryan, 238
Best-of-flightgear.dk, 78
Black Lions, 236, 237
Blue Angel Club, 20
Blue Angel Dreams 50th Anniversary, 156
Blue Suit, 171, 178-179, 187, 306-307, 311, 313-314
Blueangelphantoms.com, 71
Bonza, 176
Borgnine, Ernest, 198
Bradley, Travis J., 197, 229, 232-236, 239-240
Bratton, Julius, 161, 168, 201
California National Productions, 89
Call Sign, 152, 169, 194, 196, 198, 204-217
Campbell, Bill, 215
Cassidy, Mel, 18
Charlie, 217
Chennault Air Force Base, 9
Chewy, 169
Chief of Naval Air Training, 18, 156
Chief of Naval Operations, 97
CHOMPS, 166
CNATRA, 158
Collins, Emmet, 63
Colwell, David, 308
Combat Air Musem, 101-103
Conrad, Becky, 70-71, 92
Cormier, Zeke, 48, 71
Corpus Christi, TX, 7, 52
COVID-19, 69, 180, 232, 266
Cox, James, 166
Cragen, Jack, 227

Crew Chief, 223, 226-227, 240-244, 246-247, 250-251
Crowe, Dayle, 45
D.A.R.Y.L., 187, 215
Davidson, Ralph, 18
Davis, J. D., 121
Defense Logistics Agency, 125
Defense Supply Agency, 125
DeRitter, Brian, 10-14
DeRitter, Jill, 14
DH-151, 113-118, 120-121, 123-125, 129-130, 133, 141
Dietlin, Chelsea, 252-253
DOC, 212
Dom, George, 175, 309
Donile, Vince, 118-119
Doolittle, Jimmy, 18
Doyle, Eric, 238-239, 266
Drewelow, Kevin, 101
DuPont 6361, 152
DuPont Imron, 152
EEK-4A/P, 141, 152, 154
Eglin Air Force Base, 236
F-4J, 7, 10, 119, 121
F/A-18, 7, 16, 265-266
F11F-1, 7
F6F, 7
F8F, 7
F9F-2, 7
F9F5, 7
F9F8, 7
Fat Albert, 172, 311
Feminine Flair, 306-307
findagrave.com, 51
Flight Suits, Ltd., 142, 148, 157, 164, 269, 273
flighthelmet.com, 57, 109
Foley, John, 15-16, 181-182
Fonda, Henry, 316
Ford, Harrison, 199
Forster, Katlin, 168
***Foundation Magazine*,** 58
Fudge, 213
Fugate, Robert, 224-225, 227-228
Galloway, Jared, 242
Garland, Scott, 236
General Textile Mills, 32, 62, 65, 77, 268, 270
GENTEX, 32, 41, 49-51, 57-58, 63, 65, 74, 97, 99, 105, 107, 109, 113-115, 124, 129-130, 268-276, 278, 285-286
GENTEXITE, 51
Gershon, Mike, 144-145
Gibson & Barnes, 122, 142, 148, 151-154, 157, 256-257, 264, 269-271
Gledzinski, Hank, 96
Glenn, John, 316
Graham, Frank, 44
Griffith, Fred, 67
Grim Reapers, 236
Gucci, 15, 181-182, 214-215
Guidry, Chris, 9
Guidry, Melissa, 9
Guidry, Renée, 9
Guidry, Stephen, 9

H-1, 32, 36-38, 41-46, 48-49, 55, 65
H-2, 49, 51, 54-55, 61, 65
H-3, 28, 49, 56-57, 59-60, 62-63, 65-66, 74, 81, 90, 93
H-4, 28, 49, 56, 65-66, 74, 81, 90, 93
Haley, James, 165, 233
Hancock, Ben, 214
Hardman Tool & Engineering Co., 84, 104-105, 120, 133, 199
Harris, Art, 199
Harvey, Gerald, 242-243
Haumesser, Mary E., 22
Hawkins, Arthur, 37, 41-44, 52-56
Heitt, Chuck, 46
Helmet, General Utility, 125
Hempler, Brandon, 180
Hendler, Edwin, 33
Hendrix, Cody, 254-255 **HGU-33/P,** 115, 125, 129, 141-143, 146-149, 151, 154, 158, 162-164
HGU-55/P, 113, 147, 151-152, 154-158, 163-164, 195, 197, 199
Hiett, Chuck, 46
Hobbs, Klio, 101
Holley, Ed, 81
Holloway, Charles E. Jr., 70
Homecoming, 9, 12-14, 264-266
Hope, Bob, 316
Hoskins, George, 38
Hultgreen, Kara, 206
Hunter, Herb, 81
Hunter, Perry, 180
Hurricane Sally, 231, 234
Inherent Resolve, 236-237
IROC, 216
Jar-Head, 46
Johnson, Michael, 100
Johnson, Tara, 214
Joint Base Andrews, 13
Jones, Tom, 41
Jones, Warner, 91
Kesselring, Brian, 165-169, 240, 266
King, Larry, 316
Knotts, Carol, 198
Knuckles, 176
Korean War, 40
Kozach, Al, 114
Lawman, 214
Lehman, John, 205
Life Support, 10-12, 16, 223, 225, 227, 229-230, 232-233, 240, 250, 252, 255-256, 258
Linn, Dina, 69
Lite Manufacturing, 65
Lombard, Charles, 273-274
Loney, Hal, 101
Longworth, Bob, 38
M-1944, 28
MACE, 204
Magda, John, 37-38
Marine Corps Air Station Yuma, 204
Maslowski, Jim, 121
Maverick, 217
McCain, John, 316
McGillis, Kelly, 217
McGuire's Irish Pub, 179-180

Miller, Sam, 306-314
Mine Safety Appliance, 77
MO, 213
Moneymaker, Pat, 309
Moore, Bob, 62, 219-222
Mullins, Jeff, 158, 189-194, 253
Murphy, Mike, 243
Murphy, Pat, 41
NAF El Centro, 98, 120, 265
NAS Belle Chase, 63
NAS Corpus Christi, 7
NAS Jacksonville, 7, 18
NAS Whiting Field, 7
National Museum of Naval Aviation, 16, 46, 67, 72
Naval Aviation Museum Foundation, 318, 321
Naval Aviation News, 50, 74
Naval Training Command, 96
NAVINFOEAST, 253
NAVPERS, 59
Navy Blue Lancers, 19
Navy Office of Information, 253
Navy Personnel Command, 77
New Orleans States Newspaper, 63
New Yorker Magazine, 19-20
Newbies, 171, 175, 177-178, 180, 187
Nimitz, Chester, 18
No, 214
Nord, Mike, 148
Norris, Steve, 97, 108, 113-114, 124-125
Nothhaft, Ryan, 71-72, 117-118
O'Connor, Kieron, 309
O'Mara, Kevin, 120-124, 134
Operations Tempo, 252
OPTEMPO, 252, 256
P-1, 65-69, 71-73
Packer, Larry, 213
Parachute Rigger, 11, 218, 225, 232, 236, 238
Paramount Pictures, 204, 217
Paul Fournet Air Service, 9
Pensacola Naval Air Station, 12-14, 67, 72, 86, 118-119, 262-264, 266
Perrault, Mark, 62
Pierozzi, Nello, 83
Pittsburg Press, 22-23
Poptart, 213-214
Preston, Jim, 115, 132, 139-140
PRK-37/P, 116-120, 129, 141-142, 147, 151, 154
Proceedings Magazine, 205
Quality Assurance, 12
Rasmussen, Thomas Jefferson, 81
Reflections of Blue, 198
Reynolds, Jack, 255, 260
Rhodes, Raleigh, 19, 26, 37-38
Rich, Bud, 41
Rickoff, Cary, 167
Robb, Kevin, 237-238
Robcke, Jake, 38
Rottgering, Dave, 94, 299
Rud, Gil, 149-150, 313
Saffir-Simpson, 231
Sailor of the Year, 13
Salimbeti, Andrea, 36
San Diego Air & Space Museum, 72-73, 75, 93

Satan's Kittens, 38
Schaffer, Dan, 200
Schartz, Sheldon, 81
Schnick, Jim, 86
Scholl, Ryan, 155
Schram, Dick, 100, 124, 136-137
Schwaiger, Robert, 98
Schwartz, "Lefty", 68
Sewart Air Force Base, 52
Shipley, Dan, 176
Shultz, Dave, 136, 138
Sierra Engineering Company, 97, 99, 105, 113, 124
Simpson, Travis, 200
Slezak, John, 48, 54, 61, 68, 90
Slote & Klein, 26-27
Smithsonian Air & Space Museum, 67
Soupy, 215
Stallings, Scott, 110
Steadman, Dan, 176
Sugalski, Martha, 201
Sundowners, 222
Switzer, Bill, 121
T. T. Wentworth Museum, 16, 321
T2V, 62
Tedrow, Mark, 216
The Magic of Flight, 155
The Red Helmet, 8, 68
The Sewing Box, 306-308, 311
Timber, 183, 185
Tootsie, 166
TOP GUN, 204, 217
TOPTEX, 274-275
Trader Jon, 16, 315-318, 320-325
Traderomics, 316
U. S. Air Force, 68, 123, 125, 129, 151, 179, 190, 220, 274-276, 305
U. S. Army, 22, 85-86, 125
U. S. Navy Flight Exhibition Team, 16, 18, 22
Umstead, Skip, 134, 242
Union Bulletin, 86
USAF Thunderbirds, 8, 13, 66, 70, 120
USS George H. W. Bush, 236
Van Oosten, Lucien, 273
Ventimiglia, Duke, 86-87
Versluis, Darby, 255, 257, 260
VF-111, 222
VF-124, 110
VFA-101, 236
VFA-213, 237
VIP, 60-61, 88, 94, 112, 195-200, 202
Vonier, Dick, 94
Voris, Butch, 18-20, 32, 38, 41, 44, 50, 71-72
Votaw, Jill Hawkins, 51
Walborn, Ben, 167, 180
Wedemeyer, Scott, 309
Wegge, James, 142, 154, 157, 164
Weissmann, Martin, 315, 318, 320
Whiskers, 168
Whitehurst, Candy, 307-308, 311-312, 314
Wickendoll, Maurice, 18
Widener, Nathan, 236
Willis & Geiger, 26-27
Woods, C. J., 255, 257, 261
Wooldridge, Greg, Foreword, 187, 309

Wright Patterson AFB, 125
Young, Rick, 183-188
YouTube, 156
Zastoupil, Frank, 166
Zimmerman, Robert, 273-277

Also Available

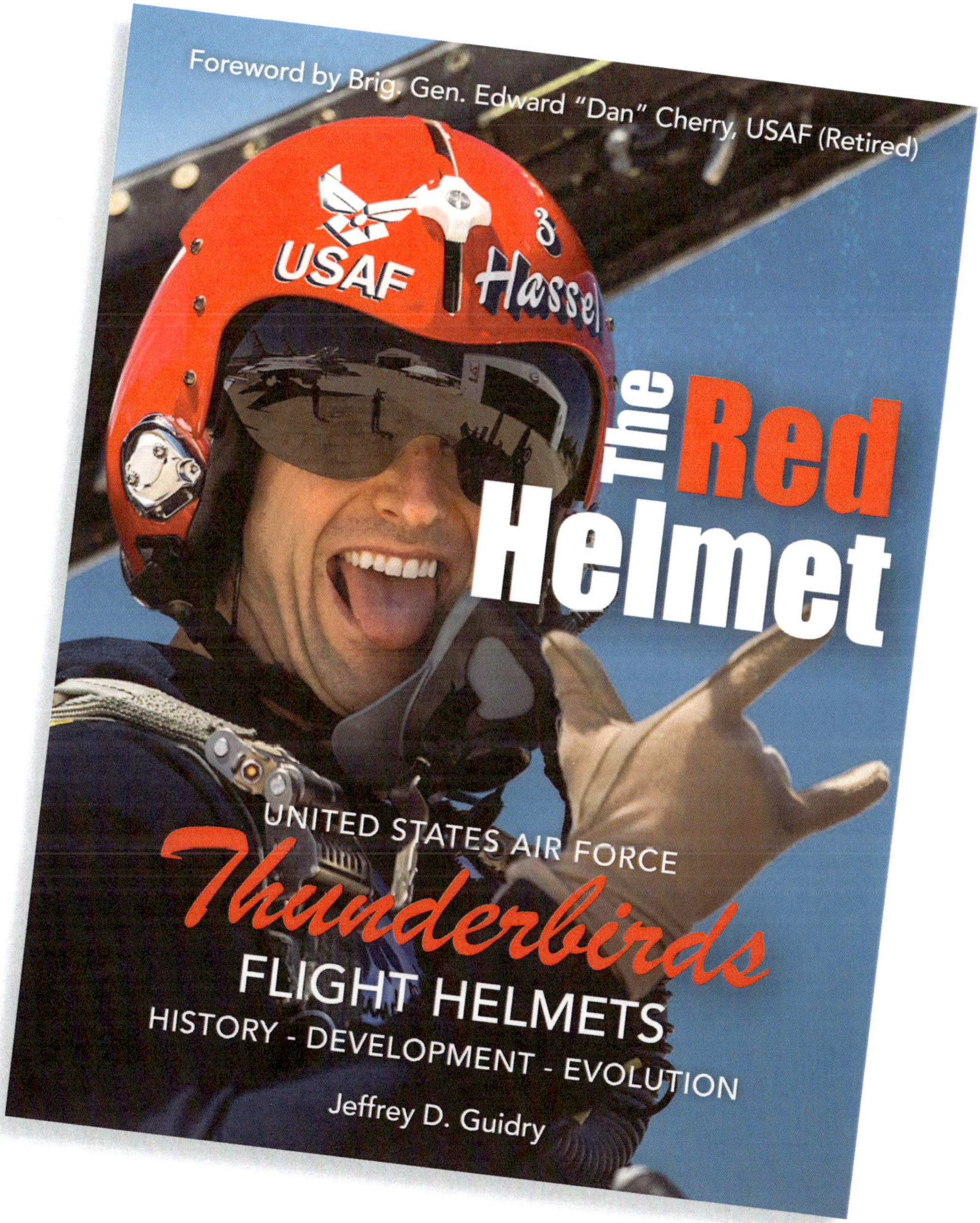

www.ingramcontent.com/pod-product-compliance
Lightning Source LLC
LaVergne TN
LVRC101546220926
848486LV00041B/384

* 9 7 8 1 9 4 3 4 9 2 9 8 5 *